About Island Press

Since 1984, the nonprofit Island Press has been stimulating, shaping, and communicating the ideas that are essential for solving environmental problems worldwide. With more than 800 titles in print and some 40 new releases each year, we are the nation's leading publisher on environmental issues. We identify innovative thinkers and emerging trends in the environmental field. We work with world-renowned experts and authors to develop cross-disciplinary solutions to environmental challenges.

Island Press designs and implements coordinated book publication campaigns in order to communicate our critical messages in print, in person, and online using the latest technologies, programs, and the media. Our goal: to reach targeted audiences—scientists, policymakers, environmental advocates, the media, and concerned citizens—who can and will take action to protect the plants and animals that enrich our world, the ecosystems we need to survive, the water we drink, and the air we breathe.

Island Press gratefully acknowledges the support of its work by the Agua Fund, Inc., The Margaret A. Cargill Foundation, Betsy and Jesse Fink Foundation, The William and Flora Hewlett Foundation, The Kresge Foundation, The Forrest and Frances Lattner Foundation, The Andrew W. Mellon Foundation, The Curtis and Edith Munson Foundation, The Overbrook Foundation, The David and Lucile Packard Foundation, The Summit Foundation, Trust for Architectural Easements, The Winslow Foundation, and other generous donors.

The opinions expressed in this book are those of the author(s) and do not necessarily reflect the views of our donors.

A complete list of titles in this series can be found in the back of this book.

The Society for Ecological Restoration (SER) is an international nonprofit organization whose mission is to promote ecological restoration as a means to sustaining the diversity of life on Earth and reestablishing an ecologically healthy relationship between nature and culture. Since its incorporation in 1988, SER has been promoting the science and practice of ecological restoration around the world through its publications, conferences, and chapters.

SER is a rapidly growing community of restoration ecologists and ecological restoration practitioners dedicated to developing science-based restoration practices around the globe. With members in more than fifty countries and all fifty US states, SER is the world's leading restoration organization. For more information or to become a member, e-mail us at info@ser.org, or visit our website at www.ser.org.

MAKING NATURE WHOLE

Making Nature Whole

A History of Ecological Restoration

William R. Jordan III and George M. Lubick

ISLANDPRESS

Washington | Covelo | London

Text design and typesetting by Karen Wenk
Printed using Electra

Library of Congress Cataloging-in-Publication Data

Jordan, William R.
  Making nature whole : a history of ecological restoration / William R. Jordan III and George M. Lubick.
      p. cm. — (The science and practice of ecological restoration)
  Includes bibliographical references and index.
  ISBN-13: 978-1-59726-512-6 (cloth : alk. paper)
  ISBN-10: 1-59726-512-8 (cloth : alk. paper)
  ISBN-13: 978-1-59726-513-3 (pbk. : alk. paper)
  ISBN-10: 1-59726-513-6 (pbk. : alk. paper) 1. Restoration ecology. I. Lubick, George M., 1943– II. Title.
  QH541.15.R45J65 2011
  333.71'53—dc22

                                                                                                2011005748

Printed on recycled, acid-free paper

Manufactured in the United States of America
10 9 8 7 6 5 4 3 2 1

Keywords: Island Press, Society for Ecological Restoration, restoration ecology, ecological restoration, environmental history, environmental philosophy, ecocentric restoration, meliorative land management, John T. Curtis Prairie

In memory of Walter G. Rosen (1930–2006)—teacher, mentor, democrat, activist, and man of the book. He coined the word *biodiversity* and was an early champion of ecological restoration.

CONTENTS

ACKNOWLEDGMENTS

Many colleagues and friends have contributed to our exploration of the history of ecological restoration, and we deeply appreciate the knowledge, advice, criticism, and support they provided. Some provided firsthand knowledge of important developments, others helped with technical and scientific matters, and some shared valuable insights into the character and development of this fascinating form of land management. We extend our warm thanks to these friends and colleagues, while acknowledging that any mistakes in our account are entirely our own responsibility.

Many of these advisors appear in the story itself. Here we list only those who helped in ways not fully acknowledged in the text: Arnie Alanen, David Amme, Kat Anderson, Dean Apostol, Nat Barrett, Mark Barrow, Eleanor Betz, Charles Beveridge, Nancy Bissett, Tom Blewett, Nancy Braker, Steve Broome, Steve Buck, J. Baird Callicott, David Channell, Stephen Christy, Andy Clewell, Ted Cochran, Frank Court, Chris Craft, Kent Curtis, Narayan Desai, Karen Dunn, Dave Egan, Jerry Emory, Edward Finn, S. J., Robin Freeman, Peter Friederici, Peter Fulé, Roger Gettig, Steve Glass, Gene Hargrove, Dave Harmon, Liam Heneghan, Ted Hiebert, Eric Higgs, Jeff Horst, Benjamin Hunnicutt, James Knight, Cynthia Kosso, James Krokar, Peter Leigh, Tein MacDonald, Dennis Martinez, Curt Meine, Steve Moddemeyer, Tim Moore, Steve Packard, Greg Powell, Steve Rauh, Kay Read, Chris Richard, Karen Rodriquez, Meg Ronsheim, Michael Rosenzweig, Leslie Sauer, Richard Sellars, Mike Shank, Jake Sigg, Tom Simpson, John Stanley, Joanne Stucjus, Susan Swisher, Bill Tischler, Dave Tomblin, Amy Tyson, Mark Wegener, Linda Sargent Wood, and Joy Zedler.

Special thanks to Barbara Jordan for checking drafts and for her helpful responses to a succession of ideas and quandaries and to William R. Jordan IV for his insightful responses to many of the ideas we explore in this book.

We gratefully acknowledge support for this project from the Bradshaw Knight Foundation, the Graham Martin Foundation, and the National Science Foundation (Award SES-0423947).

Introduction

The idea of restoration is not new. Life, as it happens, is full of accidents, mishaps, and untoward events, from breaking a tool to growing old and dying, and the desire to avoid the consequences of these events by reversing or undoing them in order to return to some happier previous or "original" condition seems to be as old as our species. It is the root aspiration behind religion, that oldest of human institutions. It is also a constant theme of both public and private life, and it may even be grounded in the capacity for self-repair that is a peculiarity of life itself.

As far as landscapes and ecosystems are concerned, restoration has a long history. The use of fire by hominids to create, maintain—and so in a sense to restore—habitat suitable for themselves predated the appearance of our species,[1] and the development of agriculture only added to the number of techniques for what might be called restorative land management, including not only practices such as shifting agriculture, crop rotation, and the manuring and fallowing of land but also reforestation and sustained-yield forestry; game, fish, and range management; and certain forms of landscape design.[2] And in the realm of imagination, where relationships are negotiated, values created, and consciences formed, it includes the invention of world renewal rituals and taboos and practices to maintain relationships with totemic species and populations of prey animals that date back as far as we have or can infer any record. The aims of these practices have generally been human centered, however, entailing restoration or maintenance of certain features of ecosystems—such as productivity, value as habitat for favored species or for the souls of departed ancestors, or other forms of natural capital—valued because they enhance an ecological system as habitat for a particular culture.

1

Early in the twentieth century, however, a handful of managers at scattered sites undertook projects aimed at restoring whole ecosystems, bringing back not just selected features of the model system but all of them. The motives behind this new and in some ways odd enterprise were complicated: a mixture of curiosity, scientific, historic, and aesthetic interest, nostalgia, and respect for the old ecosystems, together with the idea that the old ecosystems are ecologically privileged assemblages of organisms, endowed with distinctive qualities of stability, beauty, and self-organizing capacity, and so might be useful as models for human habitat.

Concern about human habitat was not new, of course. But the idea of re-creating an entire ecosystem, community, or landscape, complete with all its parts and processes, *was* a new idea. It was also an important idea, and to distinguish it from other forms of land management, which we might call meliorative land management (i.e., making an environment "better" for someone), we are calling this *ecocentric restoration*, which is restoration focused on the literal re-creation of a previously existing ecosystem, including not just some but all its parts and processes. This entails everything we do to an ecosystem or a landscape in an ongoing attempt to compensate for novel or "outside" influences on it in such a way that it continues to behave or can resume behaving as if these influences were not present.[3]

The difference between these two versions of environmental husbandry or caretaking is clearly illustrated by a story that came out of the early work at the University of Wisconsin Arboretum in Madison, where some of the earliest experiments in this new form of environmental restoration were carried out beginning in the mid-1930s. As the Civilian Conservation Corps crews working on the site completed their plantings in the area where they were restoring a tallgrass prairie, Ted Sperry, the ecologist who was supervising the project, asked them to disassemble the old stone walls at the site and rescatter the stones as they might have been left by retreating glaciers thirteen thousand years earlier.[4] One imagines the comments this must have elicited from crews of young men, many of whom had grown up on farms and had spent time picking stones out of fields to clear them for crops. But the point is that this project wasn't about crops, productivity, or the crew members' feelings. It was about the prairie.

This book begins in a sense with this story of the stones in the making of what has since been named the John T. Curtis Prairie, in honor of one of its creators. The UW–Madison project was not unique. The notion of creating faithful representations of natural landscapes and ecosystems has a pre-history dating back to the nineteenth century and even earlier, and projects very much like the one at the arboretum were under way at several locations by the early 1930s. In our research for this book we have identified six—four in the United States and two in Australia—all dating from around the same time. Taken together these projects represented, in the literalism of their attempt to re-create historic associations, a threshold in conservation practice. By undertaking the re-creation of whole ecosystems in this way, insisting on including what Aldo Leopold later called "all the parts," these managers had placed on the ground, if not the real thing exactly, at least a very provocative thing, and it has since become clear that this is a distinctive game to be playing with nature and that it has important implications for conservation.

It is this version of restoration that we explore in this book. We ask, How did those early projects come about? What did those who dreamed them up and carried them out think they were doing? Why were they doing it? And what have people made of this form of land management in the three quarters of a century since?

What we will find is, first of all, a case of arrested development. Whereas meliorative restoration has a long and more or less continuous history, one clear indication that ecocentric restoration is fundamentally different from it is that it has a much shorter history and has evoked a very different kind of response. Pretty much everyone is in favor of meliorative restoration, or "conservation," at least in principle, but this has not been true of ecocentric restoration. In fact, the early efforts at what we are calling ecocentric restoration proved curiously sterile. Most were abandoned within a few decades, and half a century passed before conservationists began to take the idea these projects represented seriously and to implement it in a serious way. Only in the past few decades, beginning in the 1970s, have land managers, conservationists, and environmentalists begun to practice this form of land management and to discover ways of realizing its distinctive value.

This ongoing process of discovery has included the realization that restoration, understood as the attempt to reverse the effects of novel influences on preexisting ecosystems, is crucial to the survival of the "natural" (more accurately historic) ecosystems that have been at the center of environmental thinking for much of the past century. Although the earliest attempts at ecocentric restoration involved the wholesale reassembly of ecological communities on drastically altered sites such as abandoned farm fields, managers have come to realize that these projects actually provide a model for the "preservation" (more accurately the perpetuation) of any ecological system subject to novel or outside influences—in other words, ultimately all of them, including "real-world" systems such as national parks.

This process has included the discovery of the value of this form of land management as a context for learning and for raising questions and testing ideas about the ecosystems being restored. Other forms of land management have heuristic value, of course. But only ecocentric restoration is focused explicitly on the ecology (and, not incidentally, the history) of the old ecosystem.

It has also included exploration of the value of ecocentric restoration as an experience that fosters the values on which attempts to reduce human abuse of ecosystems, including abuses that contribute to the prospect of anthropogenic climate change, ultimately depend.

And it has included the realization that ecocentric restoration provides the key to a positive relationship with old, often ecologically obsolete ecosystems, not only addressing but literally solving the dilemma of human use of "natural" areas that has bedeviled conservationists ever since they began setting aside parcels of land as preserves.

These benefits of ecocentric restoration are now widely recognized, thanks to a quarter of a century or more of exploration, discussion, and experimentation by a growing culture of restorationists. Less widely recognized are several others that have been brought home to us as we researched and wrote this book.

These are:

The value of ecocentric restoration as a self-conscious encounter with nature as other in the form of ecosystems that were there when we got there and owe nothing to us.[5] This is a distinctive feature of ecocentric restoration. It is an invention not of indigenous people, who have made themselves at home in an ecosystem, but of newcomers, a response to the mixture of regret, nostalgia, and curiosity some feel on looking back at the "original" landscape they and their predecessors have altered, often beyond recognition. This is something quite different from attempts to maintain land conceived as a garden, which historian Marcus Hall finds to be characteristic of restoration efforts in Europe. Historically, the idea emerged "especially in North America," according to Hall; we would say the Anglo/New World, including Australia.[6] This is important because it offers a way of confronting our experience of ecosystems and their inhabitants as other than us at the level of the ecosystem. What we have in mind here has nothing to do with the nature–culture dualism environmental thinkers have often decried as the root of our environmental crisis. Rather, it has to do with the fact that, as self-conscious creatures, humans experience the world as something they are both part of and apart from. That being the case, if the aim of environmentalism is to provide the means for negotiating a healthy relationship with the environment, then it has to provide psychologically effective ways of dealing with both aspects of this experience. From this perspective there are just two forms of land management: ecocentric restoration and all the others.

Its value as a way of making us aware of our influence on an ecosystem or landscape. All the things the restorationist does are an attempt to compensate for that influence, and these add up to an ecological definition of who we are—that is, a definition in terms of what we have done to what was here before we got here.[7]

Its value as tribute. Because including "all the parts" in most instances cannot be justified in purely self-interested terms and often entails sacrifices of immediate human interests (think poison ivy, rattlesnakes, mosquitoes[8]), it at least implies the valuation of nature for its own sake. If meliorative land management treats the land as human habitat, enacting and celebrating our membership in the land community, ecocentric restoration acknowledges the givenness, strangeness, and otherness of nature, and it may be the strongest thing we can do to acknowledge and pay tribute to its intrinsic value—its value, that is, apart from our own interests.

We are aware that many question the whole idea of disinterested action and that the question of altruism is a puzzle that has stymied both biologists and philosophers. However, it seems to us that when we find someone going to the trouble of restoring an entire ecosystem, including all its species and processes, the apparently inconsequential and even the unappealing and dangerous ones, that reads as what we will call operational altruism—that is, whatever is going on in that person's head or heart, and whatever he is saying to justify this odd behavior in utilitarian or self-interested terms, what he is doing is indistinguishable from the actions of, say, the Good Samaritan in the parable.[9]

Besides this, ecocentric restoration is important because of the questions it raises and the ambiguities it dramatizes. Every form of land management does this, but each of them raises its own distinctive questions and ambiguities. Ecocentric restoration forces the practitioner to confront questions about the proper role of humans in the life of a natural ecosystem and about whether restoration is the epitome of a responsible relationship with nature or, as some have argued, just another excuse for human aggrandizement. These questions are easy to ignore in a landscape regarded as a preserve or a garden. They are almost impossible to ignore in a landscape regarded as under restoration.[10]

As these various benefits have become clear, approaches to land management that are informed by the idea of "putting everything back" have assumed growing importance as a conservation strategy, prompting former interior secretary Bruce Babbitt to declare restoration the theme of his term as secretary of the interior (1993–2001)[11] and biologist E. O. Wilson to assert that "the next century will . . . be the era of restoration in ecology."[12]

This raises some fascinating and even urgent questions. If what we are calling ecocentric restoration has a crucial role to play in the conservation of the historic associations that have been a special concern of environmentalists at least since the time of John Muir, then why did several generations of conservationists and environmentalists variously ignore, misunderstand, and on occasion resist it? What does this tell us, not only about the strengths and limitations of ecocentric restoration but also about these various environmentalisms and their limitations? These are questions we will consider as we explore the history of this emerging form of land management.

* * *

A few comments on our approach are in order. The emergence of ecocentric restoration as a conservation strategy during the past few decades has led to a good deal of thought and discussion devoted to questions such as the aims of restoration, how best to define these aims in various situations, and the proper role of restoration in the management of areas such as the national parks that have been set aside explicitly as natural areas. For the most part, however, this discussion has emphasized the continuity—or overlap—that naturally exists between ecocentric restoration and the various forms of meliorative land management.

In this book, in contrast, we emphasize the distinction between ecocentric restoration and the various forms of meliorative land management. This is not because we regard ecocentric restoration as better than or morally—or even ecologically—superior to meliorative ecosystem management. In fact, we don't. It is certainly not that we regard the old ecosystems that commonly serve as models for restoration projects as an ideal. In fact, we see them in value-neutral terms, as historic, which is roughly the opposite of ideal. Nor is it because we are interested in purity of motives or because we are unaware that there are infinite degrees of ecocentricity and that none are entirely free of self-interest. Many forms of meliorative land management (for example, forestry or game management) entail restoration of at least some elements of a preexisting ecosystem and may be—and often are—undertaken in a spirit of profound respect for nature. By the same token, not even the "purest" attempts at ecocentric restoration are entirely free of self-interest.

At the same time, it is clear that, although it may be impossible to distinguish them clearly in practice, these two motives define very different games to be playing with nature, just as they define very different relationships with other people—roughly the difference between friendship and business. In other words, they have very different consequences not only for the ecosystems they create but also for the meanings and values they enact. Both are important, indeed crucial components of a healthy relationship between us and our environment. And that is precisely why it is important that the distinction between them not be obscured or downplayed.

This may not be of great interest to the land manager who is concerned exclusively with the ecosystem and finding the best way to keep it healthy or productive. But it is of fundamental importance to anyone who takes an interest in an old ecosystem or in land management practices as contexts for negotiating a relationship with the land and creating the meanings and values that emerge from such a relationship. This is at least as important as the purely ecological, economic, or "practical" aspects of any kind of land management. After all, the future of nature on this planet may well depend on what Aldo Leopold called "our intellectual emphasis, loyalties, affections and convictions"[13]—that is, on our values. If that is true, then the future of the natural world may ultimately depend on the full range of values generated by the various versions of land management. And in order to achieve those it will be necessary, at least in a few places, to insist on the distinctions between them.

Take the question of earthworms, for example. Ecologists have discovered that earthworms were extirpated from large areas in the upper Midwest during the most recent glaciation and that all the species found in these areas now are exotics, introduced in many cases by anglers disposing of bait.[14] Even though these newcomers affect the composition and functioning of the ecosystems they now inhabit, no one imagines eradicating them from the Midwest, or even from preserves of any significant size. Yet it might be of great value to attempt to create, say, a worm-free acre somewhere in each midwestern state—interesting to ecologists, certainly, but also as a performance of sorts, evoking reflection on ecological change, its consequences, and our role in bringing it about.

It is when we consider restoration in this way, as a context for reflection, value creation, and the making of meaning—certainly not trivial or incidental aspects of any activity—that distinctions such as that between ecocentric and meliorative restoration are especially salient. Although the world "out there" is analog, anthropologist Roy Rappaport has pointed out that the creation of meaning entails digitizing our experience of the world, as we do when, for example, we designate an entity as, say, a human, a label that overlooks the blurry edges of this category. (We debate endlessly the question of when a fertilized egg becomes a person. And as a priest recently asked Bill Jordan, "What am I to call a cadaver from which—or whom—we are harvesting organs?") To make sense—and meaning—out of this blooming, blurry confusion we digitize it, that is, impose on it sharp, either–or distinctions as a basis for the creation of meaning through language and ritual.[15] Take, for example, our observation that ecocentric restoration is uniquely a way of enacting the idea of the intrinsic value of nature at the level of the ecosystem. Our experience of altruism may be irreducibly ambiguous, as the philosophers tell us; we simply cannot know that any action, even our own, is free of self-interest. Yet the distinction between self-interested and altruistic (that is, other-regarding) behavior is central to moral experience, and for this reason it is important to have ways of acting it out in dramatic, digitized ways. It is because we are interested in the realization of the value of restoration in all the dimensions of value—not just the ecological—that we will be keeping distinctions such as this one in mind.

One more comment. We are approaching the history of ecocentric restoration as a history not merely of its "invention" but of the discovery and realization of its distinctive value.

As the Vikings' encounter with the New World in the eleventh century (or possibly that of the Chinese early in the fifteenth) illustrates, finding or doing a thing is not the same as discovering it. In either case the real event, the act of genius, is not just the doing or making or finding of a thing. It is realizing its distinctive value, articulating that value, and then deliberately exploiting it for the sake of that value.

Recognition of these distinct steps in the process of innovation and discovery is important because unless we are aware of the psychological distance between action, idea, articulation, and realization, we are likely to confuse a new idea with an old one and so fail to realize the full value of the new idea. In the case of environmental restoration, treating ecocentric restoration as essentially another form of resource conservation, enlarged perhaps by an enlightened regard for biodiversity or noneconomic species, while downplaying the moral, psychological, and conceptual differences between these two games we play with nature, diminishes the distinctive benefits and values provided by both of them.

The distinction is crucial, not only for our project but for the future of classic ecological systems on this planet. In the past, human societies have sometimes achieved a measure of stability in their relations with their environment, but this has usually been achieved only after a wave of extinctions resulting from human invasion of a new area. Repeating this pattern on a global scale would be a biological and moral catastrophe—the catastrophe environmentalists have been warning about for several generations. To avoid this it will be necessary to distinguish between ecosystems that we propose to inhabit and those we hope to maintain against the pressure of novel influences. It will also be necessary to provide productive ways of confronting, expressing, and dealing with our own deeply ambivalent experience of the world. The various ways we manage land provide the basis for both kinds of work: outer, ecological, and economic and inner, psychological, spiritual, and moral. This work, like any work, depends on distinctions. To overlook this, to conflate ecocentric restoration with the many versions of meliorative land management, ignoring, downplaying, or blurring the distinction between them, imperils the old ecosystems and the species that compose them.

As it happens, the discovery and realization of the distinctive value of ecocentric restoration and the discussion of its relationship with other forms of land management have taken place almost entirely in the past two or three decades and have been the work not of one or two prophets or pioneers but of a diverse assortment of tinkerers, aficionados, innovators, gardeners, scientists, critics, observers, and interlopers.

Theirs is the story we undertake to outline in this book.

* * *

A note on the *E* word: Throughout this book we use the word *ecosystem* to refer to the ecological systems that provide models for ecological restoration efforts. We are aware that this term is something of an anachronism when applied to work done before the introduction of the term in the 1930s. However, we need a broad term that may be understood as referring not only to an ecological system regarded as a functional unit, with an emphasis on processes rather than composition, but also to such systems regarded as communities, associations, or landscapes. In fact, in his early use of the term, Arthur Tansley defined it broadly as "the whole system, . . . including not only the organism-complex, but also the whole complex of physical factors forming what we call the environment."[16] And the word *ecosystem* has this broad sense in popular usage as well. For both of these reasons, we chose *ecosystem* as most suitable for our purposes.

Chapter 1

Deep History

Although the idea of ecocentric restoration is a recent one, having taken shape in the early decades of the past century, it has deep roots. Searching through history we do not find full-blown precedents for ecocentric restoration projects, such as the UW–Madison Arboretum's Curtis Prairie, which most would accept as a classic example of this form of land management. But we do find many of the elements—what we might call partial precedents—of this idea. And we find them not only in practices related to the care and management of land or ecosystems but also in ideas and practices arising from relationships people have formed with other humans and with their gods.

As far as human relations with the nonhuman environment are concerned, the picture that emerges from a historical overview is mixed, with humans (like any other species) bringing about changes in an ecosystem when they first invade it but then, at least in some cases, settling down to a more or less stable relationship with the altered—and to some extent "humanized"—system. Environmental historian Curt Meine, summing up the downside of this story, describes a "sobering picture of the human past" in which "human dispersal over the past 120,000 years has been accompanied by wave after wave of extinctions and other forms of environmental degradation."[1] However, this wave of losses typically subsides as a culture settles into a reasonably stable relationship with a new—and usually diminished—suite of species. And although human societies by no means manage this consistently,[2] some achieve a sustainable relationship with their environment that they may maintain for millennia and that may be characterized by high levels of biodiversity at both the community and the species level. Ecologist Fikret Berkes, who has examined the resource

11

management practices of a number of traditional cultures sympathetically without romanticizing them, notes that "where indigenous peoples have depended on local environments for the provision of resources over long periods of time, they have often developed a stake in conserving biodiversity." This entails practices that reflect what he describes as "ecosystem-like" ideas, including the idea that all the elements of their world, including plants and nonhuman animals as well as humans, are interrelated. Regarding them as members of a family, they foster and maintain them through practices such as maintaining sacred areas and refugia, institutionalizing taboos that protect selected species from exploitation, and protecting critical life history stages of exploited species.[3]

At the same time, Berkes acknowledges the limits of traditional land management practices as far as the conservation of actual species, as distinct from biodiversity in the abstract, is concerned. Even under settled conditions, he points out, these forms of land management, while maintaining qualities such as diversity and what might be called ecosystem health or integrity, typically entail both the introduction of exotic species and the extirpation and even extinction of existing species. He notes that the concerns of peoples he has studied in managing their environment are fundamentally livelihood oriented—that is, economic and social, not biological. For example, he writes, hunters and fishers behave "in the short term as 'optimal foragers' maximizing their catch per unit of effort" and so naturally pay more attention to prey or otherwise useful species than to others. Drawing on research by ethnobiologist Eugene Hunn, he points out that this livelihood interest is reflected in traditional systems for naming and classifying organisms, which typically account for useful organisms more thoroughly than others. Summing up, he concludes that "no one has ever documented a so-called traditional preservation ethic, except perhaps with sacred sites. Indigenous peoples do not have a concern necessarily with the preservation of *all* the species in their environment (and neither do most non-indigenous peoples)."[4]

In our terms, what this means is that the sustainable land management practices of traditional peoples may provide models for sustainability but are motivated not by concern for anything like "all the parts" but by a desire to shape and maintain an ecosystem as habitat for themselves. Significantly, traditional forms of land management come closest to the idea of ecocentric restoration not in what we might call the working landscape but in the ancient institution of sacred groves, which Berkes and his colleagues mention as an exception to the generalization that indigenous peoples do not characteristically concern themselves with the conservation of all the species that share their habitat. In the ancient Mediterranean region, for example, groves were set aside as sacred to the gods (50). They were used only for worship and were protected from goats (representing, we may suppose, productive use of land), which were admitted only one day a year . . . to serve as victims of sacrifice.[5] Similarly, in India groves are set aside and maintained "for their own sake." Thus in the myth of the origin of a grove in Himchal Pradesh, India, the god demands that the grove not be regarded as belonging to the king, whose responsibility it is to appoint the priests who maintain it for the sake of the god. Similarly, M. Jha and his colleagues note that *oran*, the word for the groves, comes from the Sanskrit word for "small forest" but might also have come from the Hindi word *auron*, meaning "for others, or not for one's own use," a derivation that points directly to the idea of the concern for the whole and all its parts that distinguishes ecocentric restoration from other forms of land management. They also note that maintenance of the groves sometimes entails at least minimal management, including reintroduction of native species "if need be."[6]

Ecologically, this typically results in conservation of at least some species, resulting in local hotspots of biodiversity that have on occasion served as models for modern restoration projects. In the absence of systematic efforts to compensate for their small size and for the influence of changes in the landscape around them, however, they inevitably both lose and gain species so that their species composition and overall ecology drift in time. In the absence of protocols for ecological monitoring, however, this drift is not documented in a systematic way and is presumably not even noticed.[7]

In sum, the institution of the sacred groves clearly reflects a commitment to showing respect for nature whole and for its own sake. At the same time, in the absence of ecologically informed management to compensate for novel or "outside" influences on these island ecosystems, it cannot be counted on to ensure their well-being or the survival of their resident species over the long term.

World Renewal

Significantly, however, a sacred grove, besides being an ecological system, is also a symbol—that is, a repository of meaning. This illustrates the importance of meaning in shaping the relationship between humans and their environment, and this takes us out of the dimension of the ecological and literal and into the dimension of expressive action, performance, and make-believe. A prime example is the institution of world renewal by which many traditional societies define their experience of the world and their relationship with their environment. Historian Calvin Martin writes that the Navajo, Kiowa, and Cheyenne believe that "Things tend to run down toward evil (chaos), ugliness and disharmony" (203) and that the people play a special role in maintaining or restoring order, work that they accomplish through ritual, which reenacts—and so reinstates—the "original terms of connection" (207).[8] Significantly for us, what is involved here, as in a sacred grove, is not the maintenance of a working landscape, or even human habitat—at least not in a literal, ecological, or economic sense. Rather, it is the maintenance of the mental, psychological, moral, and spiritual structures on which such maintenance depends, the structures that actually do run down in the absence of continual maintenance in a way that an ecosystem may not. Strikingly, this entails what we might call a virtual or subjective rather than literal or objective renewal. In other words, it depends on the technologies of the imagination—ritual, art, symbol, and language—rather than on actual land management practices. And it is understood and experienced not as compromising authenticity but as renewing it, as the Sioux experiences the self-mutilation of the Sun Dance or the Christian experiences the Eucharistic reenactment of the murder and resurrection of God as an encounter with the ground of being, the really real, or sacred.

This is a dimension of experience a modern, secular society tends to overlook. But it is one we will want to keep in mind as we explore the discovery of ecocentric restoration as a context for the creation of meaning and consider debates over the value, meaning, and authenticity of restored ecosystems. In fact, the long record of successful habitation by many traditional peoples who rely on the technologies of the imagination to organize their relationship with the world suggests that they are actually more effective at ensuring the survival and well-being of a functioning ecosystem than any kind of actual land management in the absence of such technologies. As technologies of value creation and conscience formation, they also provide means for reshaping values in response to changing conditions and ways of life, ensuring the adaptability on which any long-term relationship depends.

They do not, however, ensure the perpetuation of actual ecosystems and the full complement of species that compose them. To do that it is necessary to perceive the ecosystem from the perspective of an outsider in order to perceive and act *against* the current of time and change as an ecosystem responds to changes in the technologies and economies of the people who inhabit it. Anthropologist Tim Ingold points out that this is something a society cannot do as long as it experiences itself as influencing an ecosystem from within, so that they "must move with it and never against it."[9] And literary critic Raymond Williams notes that "nature has to be thought of . . . as separate from man, before any question of intervention or command, and the method and ethics of either, can arise."[10]

This may seem paradoxical. But the reason for it is straightforward. To the extent that people think of themselves as existing within their environment and lack an idea of nature as an other—that is, as everything in the world except people, or "us"—it is impossible for them to step out of the current of time and change to observe creation from the outside and to see themselves as agents acting on, shaping, altering, and perhaps damaging an environment. By the same token, it is impossible for them to think about reversing those effects, deliberately acting to compensate for or cancel their own effects on that environment in order to restore it to some condition that existed before their arrival. The reason for this is that, as long as the human relationship with the environment is "personal," as anthropologist Mary Douglas writes, it does not provide a basis for an objectifying perspective on the unfamiliar other—that is, the other whom we do not regard as kin or a member of our family.

This has important implications. If historic ecosystems exist in a world that not only is changing but is changing in response to the pressure of novel influences that we might call history, then their survival—their rescue, so to speak, from history—depends on management designed explicitly to compensate for these influences, which is a good way to define ecocentric restoration. That, however, depends on the coming together of two linked if in some ways conflicting ideas: respect for other species and elements of nature as having value in their own right, independent of human interests; and humans' awareness of themselves as in an important sense apart from and even in certain respects transcending the rest of nature.

Environmental thinkers have typically celebrated the first of these ideas and have read the second, understood as the objectification of nature, especially as it has taken shape in the West, as a kind of cultural and intellectual original sin, or fall from nature's grace—what cultural historian Morris Berman called the "disenchantment of the world" and historian Carolyn Merchant called "the death of nature."[11] This critique overlooks three aspects of this development, however. The first is that our species has been alienated from nature in various ways for as long as we have any record. This is evident in the self-conscious use of language and ritual, not to mention the representation of animals on cave walls, all of which entail self-conscious awareness of the other as distinct from the self.

The second is that, though troubling, this sense of self is natural. It is shared in varying degrees by other species and, at the level of human reflexivity, underlies what theologian Reinhold Niebuhr called "the essential homelessness of the human spirit."[12] And the third is that it has resulted in what most would agree are goods, such as religion, science, and the arts, that transcend purely material or economic goods.

Part and Apart

Mary Douglas sorts this out with great care in a discussion of the Ho-Chunk mythology of the Trickster, which depicts the Trickster as "at first unaware of himself as an integrated individual," "amorphous," "isolated, amoral and unselfconscious, clumsy, ineffectual, an animal-like buffoon," who cannot distinguish between himself and objects in the landscape and who mistakes parts of his own body for those of others. Far from seeing this as an indication of the Ho-Chunks' failure to differentiate the self, however, Douglas sees it as an expression of their "profound reflections on the whole subject of differentiation," as the Trickster organizes his view of the world, "begins to have a more consistent set of social relations and to learn hard lessons about his physical environment," and "gradually . . . learns the functions and limits of his being" (80–81).[13]

Douglas sees this myth as paralleling the modern idea that "the movement of evolution has been towards ever-increasing complexification and self-awareness" but notes that, contrary to the interpretations of some anthropologists, the earlier condition of culture represented by the fecklessness of the Trickster "is not pre-logical." It is, rather, "pre-Copernican." That is, "Its world revolves around the observer who is trying to interpret his experiences. Gradually he separates himself from his environment and perceives his real limitations and powers."

"Above all this pre-Copernican world is personal," she writes (81). And, as far as the difference between the primitive (a term Douglas defends on the grounds that rejecting it implies that *primitive* denotes an inferior condition) and ourselves is concerned,

> There is only one kind of differentiation in thought that is relevant, and that provides a criterion that we can apply equally to different cultures and to the history of our own scientific ideas. That criterion is based on the Kantian principle that thought can only advance by freeing itself from the shackles of its own subjective conditions. The first Copernican revolution, the discovery that only man's subjective viewpoint made the sun seem to revolve around the earth, is

continually renewed. In our own culture mathematics first and later logic, now history, now language and now thought processes themselves and even knowledge of the self and society, are fields of knowledge progressively freed from the subjective limitations of the mind. To the extent to which sociology, anthropology and psychology are possible in it, our own type of culture needs to be distinguished from others which lack this self-awareness and conscious reaching for objectivity. (79–80)

As we have seen, the idea of ecocentric restoration rests on or contains several such ideas, representing the "Copernican" distancing of self from other that Douglas considers an aspect of the psychological and intellectual emancipation that she sees as an ongoing process in all societies. These include the ideas of history and of self and society, which Douglas specifically mentions, and also the idea of nature as an object of study in abstract, objectifying scientific terms. All of these depend on a sense of oneself standing outside of something else, called "nature," which one can then regard and treat altruistically—that is, as an "other."

This ongoing "Copernican revolution" in human consciousness, though it may be taken as a ground for narrowing the scope of value to the human—or "us"—also provides the conceptual and psychological basis for a broadening of valuation beyond the human and the familiar to the unfamiliar other. This has important implications for the conservation of elements of nature that fall outside the boundaries of the familiar—that is, outside the family to which we belong. The sentimental notion that at least some human societies have enjoyed an all-inclusive sense of community is inconsistent with what anthropologists have to tell us about even human societies that are living "close to nature," do not think in terms of nature and culture, and might be supposed to have transcended or avoided these conceptual categories.

Thus anthropologist Signe Howell writes of "what I take to be a human predilection, namely to lay down premises for distinguishing between self and other." She notes that the Chewong of Malaysia, for example, who do not think in terms of nature and culture and do not "set humans uniquely apart from other beings," nevertheless do distinguish between "personages," who may be human or not, and nonpersonages, which may include "the outside world of other (and feared) humans, namely that of the Malays and Chinese." In other words, the Malays and the Chinese, not being "personages," as a tree or even a stone in the Chewongs' forest home may be, are quite literally insignificant others, and Howell comments that "unlike members of various western ecological movements, the Chewong would not accept that human beings have some *a priori* moral responsibility towards other such living beings."[14]

More broadly, anthropologist Roy Ellen writes that "nature is always constructed by reference to the human domain, and is in the last instance informed by ideas and practices concerning 'self' and 'otherness.'" As an example, he notes that the Nuaulu of Indonesia distinguish between animals and other elements of the environment that are "of the village" or "not of the village" or a "culture of the beyond"—a category that, he suggests, at times comes close to the Western idea of wilderness.[15] And philosopher Peter Singer notes that tribal moralities are often based on appeals that are effective within one's society but unacceptable outside it. In tribal societies, he writes, "Obligations are limited to members of the tribe; strangers have very limited rights, or no rights at all. Killing a member of the tribe is wrong and will be punished, but killing a member of another tribe whose path you happen to cross is laudable."[16]

So it is that the Crow tribesman, at the climax of the harrowing Sun Dance ritual of world renewal, "might cut off the joint of a finger, praying, 'O, Sun, I give this to you; send me visions and give me an enemy!'"[17] Or, to take an example from the opposite side of the continent, ethnologist William Fenton writes that warfare was deeply rooted in the culture of the Iroquois, providing the context for achieving manhood, defining status, and taking captives to sustain the population against losses. "Warfare," he writes, "was embedded in mythology, it drew strength from the sun, and it enjoyed the sanction of ritual."[18]

This means that the notion that traditional, premodern or non-Western peoples characteristically reflect a universal biocentrism or ecocentrism is dubious at best. Their families may have included members of other species, but often excluded members of their own species. And this being the case, the question for us, on our way to the inclusive valuation of other species dramatized in the act of ecocentric restoration, is, how does a society come to transcend this apparently natural anthropocentrism, ethnocentrism, or self-centrism and to recognize or confer value not only on the members of the community made valuable by their familiarity but also on the unfamiliar other?

Singer argues that this "shift from a point of view that is disinterested between individuals within a group, but not between groups, to a point of view that is fully universal" is "a tremendous change—so tremendous, in fact, that it is only just beginning to be accepted on the level of ethical reasoning and is still a long way from acceptance on the level of practice."

He also argues that this extension of value can be achieved by the application of reason alone[19]—that is, by the very process of abstraction and objectification environmental thinkers have often deplored when it results in shifts in perspective such as those represented by the rationalism of the ancient Greeks or the Scientific Revolution. Indeed, the expansion toward all-inclusiveness that restorationist George Gann has called the "logic of restoration" is logical only for a restorationist who, informed by these objectifying ideas, has stepped onto what philosopher Peter Singer has likened to the "escalator" of reason and is being carried to the top floor.

The Unfamiliar Other

With this in mind, we can proceed to a consideration of how this top-down discovery of value found expression, leading to the enlarged sense of moral enfranchisement that underlies, among many other things, the notion of ecocentric restoration. Although this is an instance of the "Copernican revolution" that Mary Douglas regards as going on in all societies, because we are concerned with the development of a form of land management that seems to have taken shape in the Anglo/New World—the United States, Canada, and Australia—we will consider these developments only in the West.

One example—or symptom—of this development was a deepening capacity for cultural criticism, represented by the Hebrew prophets coming down from the mountaintop with their jeremiads denouncing the people for their betrayal of the covenant with their distant God—a message and a tone of voice that, literary critic Herbert Schneidau suggests, depends on a degree of "alienation," which he argues reflected the Hebrews' idea of a divinity who transcends nature. This, Schneidau suggests, would be inconceivable in a society for which the community was the ultimate repository of value.[20] As he points out, however, it is just this critical perspective that has provided social critics from Isaiah and Jeremiah down to Aldo Leopold, Ed Abbey, and Barry Lopez with both a vocabulary and a place to stand when delivering their critiques of their own society.[21]

Another result of this "escalating," or extension of value to unfamiliar others, as it developed in the West, is the emergence of the idea of equality in the social and political sense—the belief that things can be alike and that to the extent they are alike should be treated the same. Because in reality no two things—and certainly no two people or cultures or species—are exactly alike, or equal, this (as the word *equal* indicates) is an abstract, mathematical idea that discounts the manifold differences between actual things.[22]

What about nature? Here the notion, famously articulated by historian Lynn White Jr., that this sense of transcendence underlies and in large part accounts for modern alienation from and abuse of nature[23] has been complicated, if not disposed of by a generation of thinkers who have made the case for a strong regard for nature in the societies that have emerged from Judeo-Christian and Greco-Roman foundations. To take just one example that has special relevance for us, Old Testament scholar Theodore Hiebert finds in the biblical tradition, and especially in the J material in the Old Testament, a warrant not for the narrowing of the horizon of value but for an actual "expansion, both in the meaning of land for Israel itself and in the spatial sweep of ancient Near Eastern geography brought into the epic vision."[24]

One manifestation of this expanding circle of valuation was the development of the principle of *bal tashchit*, meaning "do not destroy," which first appeared in Deuteronomy as an injunction to the army not to destroy trees in warfare "simply in order to render them useless to the enemy should he win." Initially, this injunction seems to have been based partly on economic considerations: An army could cut down non–fruit-bearing trees to acquire material for the construction of siegeworks, for example, and a fruit tree might be cut down if the value of its wood exceeded that of its fruit. Beyond that, however, as philosopher Eric Katz writes, the principle of *bal tashchit* insisted on the value of nature quite apart from human concerns, because it is the creation—and ultimately the possession—of God. The key element here, he writes, was the theocentrism at the heart of Hebrew thought, noting that

> This position easily renders insignificant the economic or utilitarian justifications for *bal tashchit*. The principle is not designed to make life better for humanity; it is not meant to insure a healthy and productive environment for human beings. In the terminology of environmental philosophy, it is not an *anthropocentric* principle at all; its purpose is not to guarantee or promote human interests. The purpose of *bal tashchit* is to maintain respect for God's creation.[25]

Mice and Fleas

A second reflection of this development in Judaism is the principle of *hesed*, basically a principle of hospitality entailing obligations of mercy and care for others. According to archeologist and rabbi Nelson Glueck, *hesed* seems to have appeared first as an idea of duty to family but was gradually extended beyond bloodlines to enjoin duties to all humankind, conceived as creatures of a single, transcendent God. Significantly for us, this principle is not based on feelings of affection or kindness but is an obligation imposed from above, which explicitly transcends ties based on familiarity, preference, taste, or self-interest. What it asserts is an obligation to practice what we have called operational, practical, or objective altruism — that is, action in behalf of another, whether it is accompanied by altruistic or loving feelings or not. This is dramatized in the story of Lot (Genesis 19), who was prepared to sacrifice his daughters' virginity rather than compromise his duties as host to two visiting strangers. Here, the obligations of hospitality trump those of family and affection, defining a new dispensation in which water is thicker than blood.[26]

Another reflection of the extension of value, in this case not only to the stranger but to "useless" and even dangerous creatures, occurs in the story of Noah, builder of the first lifeboat of creation and arguably the first restorationist to appear in the Judeo-Christian tradition. In an essay on hospitality, critic Jacques Derrida emphasizes that to mean anything, hospitality (which, by the way, he regards as impossible) must be undiscriminating and all-inclusive:

> Hospitality, therefore — if there is any — must, would have to, open itself to an other that is not mine, my hôte, my other, not even my neighbor or my brother, perhaps an "animal" — I do say animal, for we have to return to what one calls an animal, first of all with regards to Noah who, on God's order and until the day of peace's return, extended hospitality to animals sheltered and saved on the ark.[27]

Yet a fourth development relevant to the idea of ecocentric restoration in the biblical tradition is the institution of the Sabbath, the injunction to desist from work on the seventh day of the week, which philosopher and theologian Norman Wirzba describes as a setting aside of human desires in deference to the creation and acknowledgment of its inherent goodness apart from human interests. Significantly for us, the rule of Sabbath extended to the land, requiring fallowing of land every seventh year and dedication of its production during that time to the poor *and to wild animals* (Exodus 23:10–11). It is also significant that the Sabbath takes up only a small fraction of the week, the rest of which is devoted to productive work. Yet Wirzba argues that in this self-conscious act of awareness, deference, respect, and gratitude, creation is both completed and most fully realized.[28] In the same way, ecocentric restoration does not directly provide for people's material needs and so can never occupy a large share of the time or space available, yet it arguably represents the deepest acknowledgment of the value of the ecosystems being restored.

It is worth noting that the value of the Sabbath arises from the way it "digitizes" what is in ordinary experience a continuum of work and leisure, distinguishing them sharply—and arbitrarily—for the purpose of increasing awareness and creating meaning. This is the same point we have made regarding the importance of insisting on the distinction between ecocentric and meliorative restoration when they are being realized as occasions for the creation of meaning. In fact, in purely practical and ecological terms, work and leisure overlap broadly and sometimes even coincide. They have very different meanings, however, and this is something we will want to keep in mind later when we consider the development of restoration as a context for the creation of meaning.

Notice, too, that as an injunction to desist from work, the rule of the Sabbath in its original form provides a precedent or model not for restoration but for preservation: the deliberate setting aside of work in order to experience creation simply as given. Interestingly for us, however, Wirzba notes that Jesus violated this principle by performing cures and exorcisms on the Sabbath, providing a biblical precedent for the more active sign of respect required, we now realize, for the actual perpetuation of species and ecosystems. Here the deference at the core of the Sabbath is expressed not merely by deliberately desisting from work but by a stilling of the will in an *act* that entails obedience to an external rule or model—arguably an even stronger act of respect.[29]

During the same period, paralleling these developments by the Hebrews, Greek thought moved in a similar direction, a development that is all the more striking because it took place under different cultural conditions and reflected a different set of ideas about nature, the gods, and the place of humans in the scheme of things. Both moved beyond an ethic based on familiarity to one that enlarges the community of valued subjects beyond the familiar "us" or cohort of personages. And in both cases this was the result of an appeal to a transcendent principle: Yahweh for the Hebrews, reason for the Greeks. The Greeks' shift from *nomos* (custom) to *phusis* (nature, or "physics") paralleled the Hebrew shift from custom to God as a foundation for ethical thinking, and both were crucial steps toward the development of a way of seeing the world and valuing others beyond the familiar—or, we may say, the community.[30]

Stoicism, for example, provides a case study in Singer's model of the extension of value through the application of reason. Beginning with the work of Zeno in the closing years of the fourth century BCE, the Stoics developed a broadened, cosmopolitan conception of humanity based on the premise that the universe is rationally ordered and so is amenable to rational analysis, and that humans are both integrally part of nature and allied with all other people through possession of this essential capacity. Plutarch later praised Zeno for urging "that all the inhabitants of this world of ours should not live differentiated by their respective rules of justice into separate cities and communities, but that we should consider all men to be of one community and one polity." He also wrote that Alexander "desired that all men be subject to 'one law of reason, and one form of government and to reveal all men as one people.'"[31]

This cosmopolitan idea of value underlay, for example, the creation of a system of Roman law that granted some of the rights of citizenship to residents of the empire who were not Roman citizens. This included the creation in the third century BCE of the peregrine (that is, migratory) praetorship, a body of magistrates who dealt specifically with legal cases involving noncitizens, who until that time had been without rights under Roman law (103ff). Emphasizing universal value over national affiliation, this foreign magistrate contributed to the development of what historian Michael Grant calls "one of the most potent and effective ideas that the Romans ever originated . . . the 'law of nations' (*ius gentium*)," which defined the portions of law that applied to citizens and noncitizens alike. This law, which Grant describes as "the most civilized ideal, for practical purposes, of living that the world had ever seen," was eventually elevated into the philosophical idea of natural law, a body of precepts based on the idea of universal brotherhood and held to be valid everywhere in the world.[32]

The development of Christianity, roughly coinciding with the emergence of Imperial Rome, entailed a confluence of Jewish and Greco-Roman ideas and continued the development of an increasingly cosmopolitan conception of value. Thus St. Paul, confronting the question of how the redemptive events recorded in the Gospels applied to the Gentiles—the "others" living outside the Jewish tradition—argued that, in the aftermath of the saving act of a transcendent God, the Gentiles "become children of Abraham by God's grace through faith and not by observing the Law of the Judean nation." In other words, they are saved with their identity as Gentiles—that is, as others—intact, "without becoming Judeans . . . without being circumcised and keeping the Judean Law."[33]

This philosophy would find expression in Jesus's psychologically challenging injunction to love one's enemies and in the ideal of charity represented in the parable of the Good Samaritan (Luke 10:25–37). It would also underpin the cultural syncretism that enabled Christianity to absorb elements of many of the pagan cultures it encountered.[34] Although these principles pertained primarily to humans, some thinkers also applied them to nonhuman nature. Thus St. Augustine anticipated Aldo Leopold's injunction to "keep every cog and wheel" by 1,500 years, writing in *The City of God* that "it is not nature as seen in the light of our own convenience or inconvenience, but nature seen in her own right that gives glory to her maker." Humans, he continued, are but threads in a broader fabric and incapable of perceiving the whole, which incorporates the details that displease us "as neatly and prettily as need be." These "good gifts of nature" should not be misused, he warned, lest the abusers never receive "the better gifts of heaven."[35] And, getting down to specifics, he cautioned against the danger of abolishing "sentient beings . . . from nature altogether, whether in ignorance of the place they hold in nature, or, though we know, sacrificing them to our own convenience." "Who," he continued, "would not rather have bread in his house than mice, gold than fleas?" And, arguing for a form of stewardship that transcends human interests, he noted that "the reason of one contemplating nature prompts very different judgments from those dictated by the necessity of the needy or the desire of the voluptuous; for the former considers what value a thing in itself has in the scale of creation while necessity considers how it meets its need; reason looks for what the mental light will judge to be true, while pleasure looks for what pleasantly titillates the bodily senses."[36]

Chapter 2

Run-Up

Augustine's defense of mice and fleas was, as it remains, a minority position. At times—on the Sabbath, so to speak—people may tolerate, and even appreciate, mice in the kitchen, but the rest of the time they quite properly go about the business of excluding them. No less a champion of nature than John Muir, when not living as a mendicant out in the woods, was a prosperous orchardist, managing a business that specifically entails excluding—or killing—mice (actually, in Muir's case, ground squirrels and a host of other orchard pests).[1] And if Augustine's doctrine of regard for vermin anticipated Muir's off-hours deep ecology by a couple of millennia, the utilitarian philosophy espoused by Muir's nemesis, Gifford Pinchot, has always been well represented.

Although earlier schools of thought emphasized Western indifference to nature except as a resource, modern scholarship suggests that the idea of continuous creation, which combines a deep appreciation of nature with the idea that humans have a key role to play, not only in managing nature but in fulfilling its potential, has been a predominant feature of Western thought over the past two millennia.[2] Thus, St. Ambrose (339–397) described man "as a farmer improving the earth in partnership with God." His contemporary in the eastern church, Gregory of Nyssa (born ca. 335), although he subordinated earthly things to God, affirmed the value of nature and celebrated the arts as a way of elevating both humans and nature.[3]

Such ideas, pervasive throughout medieval society, were fully elaborated by the monastic orders that emerged early in the Christian era. Dynamic and progressive with respect to nature, this thinking arguably underlies much of the conservation thinking of our own time, providing a foundation for meliorative if not necessarily for ecocentric forms of land management. The Benedictine rule, which guided the practice of this important order from its founding in the sixth century, rested on the second chapter of the Book of Genesis, especially the passages that place humans in the Garden of Eden—not as its masters but rather in a spirit of stewardship. Following St. Benedict's (480–547) teaching that it was their duty to work as partners of God in improving his creation and re-creating Paradise out of the chaos of a fallen wilderness, the Benedictines integrated work in the fields into a life of prayer, cultivating an ethic of land stewardship. And eleven centuries before Aldo Leopold wrote of using the plow, cow, and ax to reverse environmental degradation, Irish philosopher John Scotus Erigena (815–877) first articulated the idea that the useful arts are divinely inspired pathways to salvation.[4]

All this may seem a bit remote from practical affairs. But the work (and land) ethic exemplified by the Benedictine rule implied a valuing of nature as an indispensable partner of humans not only in the Fall but in recovery from the Fall—that is, in the work that dealt with the ultimate forms of value as conceived by Christianity. In contrast with the notion of the ontological inferiority of nature that intellectuals had inherited from the Greeks, this tradition reflected a positive reading of nature. However, its aim was the recovery of Eden, understood as both the original—and so ideal—human habitat and a world under human control.

Others asserted a more altruistic vision against the prevailing, essentially anthropocentric—or theocentric—view of nature, however. St. Francis (1182–1226) taught (famously, even to birds) that nonhuman life has its own dignity and exists for its own purposes and in its own right,[5] a radical extension of Jesus's injunction to love everyone, including our enemies. Of special interest for us, at least one observer has suggested that Francis's radical egalitarianism was grounded not in the prospect of a mutually beneficial relationship with all creatures but in a spirit of poverty that underlay a radical deference to the interests of other creatures, even when, like Augustine's fleas, their interests are contrary to our own.[6]

Although this was, and has remained, a minority view, it kept cropping up. Ecologist and theologian Susan Bratton reminds us that Francis represented the ultimate expression of a tradition that had evolved over centuries and was grounded in a regard for nature that is deeply embedded in biblical tradition.[7] Taken together, the Benedictines' energetic stewardship ethic and the Franciscans' radical respect for nature are indispensable, complementary elements in the complex of ideas underlying the idea of ecocentric restoration.

A New Dignity to Nature

Paralleling in some ways the Christian idea of the role of humans in the redemption of a fallen world was the Jewish idea of *Tikkun*, or "repairing the world." Developed by Jewish mystic Isaac Luria (1534–1572), this was the idea that the creation had been interrupted by an influx of evil and that it is the responsibility of humans, despite their own fallen nature, to repair this radical flaw and bring about the perfection of creation.[8]

By Luria's time, what would later be called the Scientific Revolution was getting under way in Europe and over the next two centuries brought about the rapid development of the analytic, mechanistic idea of nature that had been taking shape in the West since the time of the ancient Greeks. The development of this objectifying perspective and its importance for the development of ecocentric restoration is well illustrated by the development of the idea that perception entails what the Aristotelian scholar F. Edward Cranz calls disjunctive thinking, as opposed to conjunctive thinking, in which the object perceived is understood to be internalized and in a crucial sense united with the knower. Cranz finds the earliest indication of this development in the writings of philosopher–theologians Anselm of Canterbury (d. 1109) and Peter Abelard (1079–1142) who proposed that an image in the mind is no *thing* at all, and is fundamentally different from the thing perceived. "What Cranz may have revealed" here, Samuel Y. Edgerton writes,

> is the crucial turning point in Western cultural history when philosophers first understood themselves as detached from nature, as outside observers limited by the inadequacy of their mental *formae* to perceiving and describing phenomena only metaphorically. At this moment, it seems, medieval peoples of western Europe commenced to understand that they were no longer living in an enchanted world where natural and supernatural forces indiscriminately confuse. . . . [Here] Anselm may have expressed for the first time the alienation of his subject "eye" from the all-seeing "gaze" of the world-God.[9]

Though deplored by many environmental thinkers as a fall from grace and primal communion with nature,[10] this new threshold of self-awareness nevertheless led to several closely related developments crucial to the idea of ecocentric restoration. The first of these was the development of perspective, a method of graphic depiction based on the idea of abstract, geometric space. This was an application to earthly subjects of a technique for depicting heavenly bodies in two dimensions that dated back to ancient times. That made it possible to depict objects "from nature" accurately in two dimensions in a way that, Edgerton argues, is consistent with human perception independent of culture. This literal, objective, and universally readable method of depiction extended the vocabulary of visual representation to those unfamiliar with the graphic conventions and idiom of a particular culture in much the same way the peregrine praetorship or the principle of *hesed* had extended the valuing of others beyond the boundaries of the familiar. At the same time, it was a step toward the commitment to realistic, point-by-point representation that underlies ecocentric restoration.

In addition, by directing attention toward the causes of phenomena understood to be inherent in nature, the new sensibility encouraged closer attention to nature and even greater respect for it, treating, as philosopher and historian R. G. Collingwood put it, "her lightest word as deserving of attention and respect."[11] A good example is astronomer Johannes Kepler's (1571–1630) discovery that the orbits of the planets are elliptical. This was a scandal for Kepler, who was committed to the idea that the heavens, being governed by divine intelligence, must move in ideal, circular patterns. He felt compelled to accept it, however, even though the observations available to him deviated from those consistent with a circular orbit by only 8 minutes of arc.[12] Another example of the emerging commitment to this kind of intellectual discipline in the observation and interpretation of nature was the flourishing in the seventeenth and eighteenth centuries of natural theology, the investigation of nature understood as a source of revelation, which contributed to the development of Darwin's theory of evolution. Most generally, the new regard for nature is evident in the deep insights into phenomena from human behavior to black holes achieved by scientists over the past few centuries.

It also contributed to the naturalizing of humans, replacing the idea that humans are in some sense not of the world with the idea that they are fully natural. As Francis Bacon put it, "artificial things do not differ from natural ones in form or essence, but only in efficient cause."[13] This has important implications for us because it meant that technology and art were not merely the imitation of nature but a way of coming to understand nature by participating in its creative processes. Thus William Eamon writes,

> Much of (Francis) Bacon's scientific program was predicated on the assumption that by imitating nature, we come closer to understanding natural processes. That is why in Solomon's House, Bacon's

utopian research institute, projects involving the imitation of natural processes were given high priority. There the researchers have "artificial wells and fountains made in imitation of the natural sources and baths"; houses where they "imitate and demonstrate meteors"; . . . furnaces that "have heats in imitation of the sun's and heavenly bodies' heats." . . . Since nature's modes of operation were no different from those of art (except in efficient cause), reproducing nature's effects by artificial means was a guarantee of man's knowledge of nature.[14]

This not only authenticates the artifact, such as a restored ecosystem; it also provides a conceptual foundation for the development of restoration ecology, that is, the exploitation of restoration as a technique for basic research, a way of raising questions and testing ideas about the ecosystems being restored.

Recovering Eden

Meanwhile, as scholars in Frauenburg, Prague, Pisa, and Cambridge took the world apart conceptually, others pursued the age-old attempt to put it back together in the most practical of ways, in the garden. The discovery of the New World prompted the founding of botanic gardens, the best known being at Padua, Leyden, Montpellier, Oxford, Paris, and Uppsala. Because their origins were academic, the gardens took on the functions of an encyclopedia, displaying in rationally designed plots the eagerly awaited specimens from exotic places. Although botanic gardens suggest a scientific purpose, they began as attempts to recreate the earthly paradise or Garden of Eden. The early explorers had thought they might actually find the lost Eden. Indeed, Christopher Columbus, skirting the coast of Venezuela on his third voyage to the Americas, wrote in his journal, apparently quite seriously, that he believed he was approaching "the earthly paradise."[15]

When explorers ultimately failed to find the Garden of Eden in the West—or East—Indies, Europeans turned their thoughts to re-creating Eden artificially. The idea was that nature had been broken as a result of the Fall and that "by gathering the scattered pieces of the jigsaw together in one place," John Prest writes, gardeners could create an "encyclopaedia of creation, just like the first Garden of Eden had been."[16]

The idea behind the botanic garden was meliorative, not only as a source of medicinal herbs but also as the ultimate melioration of recovering the ideal human habitat and, along with it, humans' understanding of and power over nature. In this sense, it was like the world renewal rituals undertaken to reconstitute what Tim Ingold characterizes as an original Distant Time, or Dreaming.[17] Gathering plants and animals together in one place allowed people to identify their useful properties and also to name them, reasserting the authority Adam had exercised in naming the creatures in the Garden. But it included hints of regard for nature for its own sake as well. The widespread collecting itself reflected the idea that all creatures had value, even if it was not yet appreciated, and this idea was reinforced by the idea that all creatures were essential links in a great chain of being that extended from God through animals and plants to inert matter, so that creation itself would fail if any one of these was lost and the chain broken.[18] (This idea, which precluded the ideas of both extinction and evolution, prevailed in the West through much of the nineteenth century.[19])

Another expression of the valuation of the unfamiliar other was the *bosco*, or little forest, an untended bit of ground some Italian gardeners left in Renaissance gardens as a trope for unaltered nature, which John Hansen Mitchell suggests represented the "invention of wilderness."[20] In any event these gestures of species cosmopolitanism are significant. Before meliorative land management could become ecocentric restoration, it would be necessary to admit all creatures, including the snake, back into the garden.

Botanic gardens were small at first, but as restoration did centuries later, they eventually expanded beyond their institutional settings to inform land management practices on a significant scale. By the end of the seventeenth century the Whig aristocracy favored gardens of "grandeur and recreation." As a result, gardening made a transition to landscape, and, according to Prest, "a great sea-change in man's outlook got underway." People gradually acknowledged the diversity of the natural world and stressed the idea that nature liked variety.[21] Although some of these designed landscapes were formal and symmetrical, English designers eventually broke with continental models and evolved their own style, inspired by an affinity for nature and influenced by the works of contemporary landscape painters, notably Claude Lorraine, Nicolas Poussin, and Salvatore Rosa. Yet, although Edward Hyams suggests that this was all in response to "a need in the English soul for what was natural," he also points out that, significantly for us, the aim of these designers was never to recreate "unimproved" or "natural" landscapes but rather to restore "an imaginary pristine perfection to nature."[22] This was the beginning of a naturalizing school of landscape design that developed through the nineteenth and into the twentieth century and eventually played an important role in the development of ecocentric restoration.

Another element of the idea of restoration that began to take shape during this period was the realization that human activities could actually degrade the environment in significant ways. Beginning with the realization, dating back to Plato,[23] that the Mediterranean region had undergone dramatic changes as a result of deforestation and grazing,[24] this led to the realization that the same thing was happening in the present, not in western Europe or North America but in a few remote tropical places, principally either actual or ecological islands, where the effects of changes in land use occurred rapidly and were especially conspicuous. Richard Grove argues that by the late seventeenth century, the resulting "colonial environmentalism," together with the professionalization of science and an emerging global network of botanical information flow, established an institutional basis for environmentalism.[25]

This growing awareness of change in nature, and of the role humans often play in it, has a long history in the United States, dating back to colonial times, and clearly reflects the tendency of European newcomers to regard the Americas as an unspoiled Eden. A notable early example was Congregationalist theologian Jonathan Edwards (1703–1758), who offered his Massachusetts congregation at Northampton Church sermons praising New England's natural beauty. In one of his later writings, "The Beauty of the World" (1758), he argued that indigenous wildlife and vegetation far surpassed the art of humans because they exhibited spiritual beauties that had been corrupted in humans as a result of the Fall.[26] A few years later, William Bartram, a pious Quaker who traveled through Georgia, the Carolinas, and Florida in the mid-1770s, described the region, then largely unsettled by Europeans, as "a terrestrial paradise" suggesting "an idea of the first appearance of the earth to man at the creation." His descriptions connected his generation with the idea of cultural primitivism, a version of what theologians Richard T. Hughes and C. Leonard Allen have called the myth of the restoration of first times. This is the idea that redemption is achieved by a return to some original condition, often construed as Eden or the society of the early Christian Church. Versions of this idea appear often in Puritan writing and, Hughes and Allen argue, have played an important role in American thinking down to the present.[27]

The same idea is reflected in the work of nineteenth-century artists such as John James Audubon, Thomas Cole, and Frederic Church, who often painted edenic, autumnal views of the natural world, suggesting twilight and expressing apprehension about the demise of such pristine landscapes as a result of encroaching civilization.[28] Audubon expressed similar regrets in his writings, suggesting that the work of artists and writers such as Washington Irving and James Fenimore Cooper might best represent the primeval landscape of early America once it was gone.[29] Prominent among these writers, though virtually unknown at the time of Audubon's death in 1851, was New England surveyor, pencil-maker, and essayist Henry David Thoreau. Expressions of concern and regret over what he perceived as the "tamed and, as it were, emasculated landscape" of post-settlement New England occur twice in his voluminous journals.[30] And though fully aware of the importance of ecosystems such as forests as economic resources, Thoreau appealed also to a higher law that he felt governed the relations between humans and nature. He questioned the primacy of humans in the scheme of nature, noting in *Walden* that "the hare in its extremity cries like a child," adding sternly, "I warn you, mothers, that my sympathies do not always make the usual phil-*anthropic* distinctions."[31] "Every creature is better alive than dead, men, moose and pine trees," he observed, "and he who understands it aright will rather preserve its life than destroy it."[32]

Ever the practical-minded Yankee for all his Transcendentalism, Thoreau not only speculated about but proposed and, in a few cases, undertook projects on behalf of the preservation, recovery, and in at least one instance actual re-creation of pieces of natural landscape. Late in his life, he offered two proposals for public parks. One called for preservation of a stretch of the Concord River bank as a public walk. The other advocated a primitive forest of 500 to 1,000 acres for every town as a common possession forever, where, as in a sacred grove, "not a stick should ever be cut for fuel." All Walden Woods, he thought, "might be preserved for our park forever, with Walden in its midst."[33]

In 1857, noting in his journal that in a young country such as the United States people had not learned the consequences of cutting down the forests, he added, "One day they will be planted, methinks, and nature reinstated to some extent."[34] And Thoreau actually undertook a bit of restoration in the spring of 1859, when he and a colleague set out some four hundred small pines, placing them "diamondwise" over 2 acres to promote recovery of a "developed" site, his bean field in Walden Woods, and speculated in his journal on how the trees would affect the habitat, discouraging field sparrows and favoring the return of thrushes.[35] And indeed his extensive observations on forest change led eventually to his writings on succession and the distribution of seeds, important contributions to natural history that indicate that Thoreau was moving toward an ecological understanding of nature.[36] His comment, in "The Dispersion of Seeds," that "when we experiment in planting forests, we find ourselves at last doing as nature does"[37] is, like Bacon's Solomon's House, an early expression of the idea that one of the most effective ways to investigate a phenomenon is to attempt to re-create it.

Thoreau's planting was never extensive. Indeed, like his experience of "wilderness" in what was virtually a suburb of Boston, it was marginal. Ecologist Daniel Botkin notes that Thoreau seems to have been content with only bits and pieces of wild nature as occasions for spiritual renewal, and he suggests that this provides the basis for a socially friendly environmentalism.[38] This is also consistent with a ritualizing sensibility that contrasts with John Muir's lifelong search for literal wilderness: Thoreau's experience suggests the possibility of redemption through imagination, whether working in a make-believe "wilderness" such as Walden Woods or a concocted one such as a restored prairie.[39]

Important as these reflections were as cultural preparation, the decisive account of the degradation of land by humans, in Europe and the United States, was the work not of an artist or philosopher but of lawyer, politician, and diplomat George Perkins Marsh, whose densely documented *Man and Nature*, published in 1864, is widely regarded as the foundation for the modern environmental movement.[40]

Marsh's work is important for us in several ways. For one thing, it brought home the lessons about human degradation of natural landscapes that had previously been associated only with remote islands. His emphasis on the value of the old associations, together with his insistence that the changes he documented actually were degradation and not merely natural deterioration, provided a foundation for the emphasis on restoration as repair of damage that historian Marcus Hall sees as distinguishing American from European approaches to environmental restoration.

Also relevant for us is Marsh's moral emphasis, his idea that sober, principled, "conserving" husbandry of natural resources would benefit both human souls and the land. Significantly, too, Marsh's thinking entailed a definite human exceptionalism that saw humans as in some ways quite distinct from nature: "Nothing," he wrote to his publisher regarding the title of his book, "is further from my belief than that man is a 'part of nature.' . . . In fact a leading object of the book is to enforce the opposite opinion." As Hall notes, citing Raymond Williams, recognition of the human ability to both destroy and restore nature made humans "more God-like and less a part of the natural world they manipulated."[41]

Ultimately, Marsh's thinking was an important step in the objectification of nature essential to ecocentric restoration and reflected little of the sympathetic valuation of nature found in the writing of earlier observers such as Edwards and Thoreau. Clearly, Marsh was a pioneer in thinking about restoration, as Hall points out.[42] But the restoration he pioneered was meliorative and not ecocentric restoration.

The polarity of perspective evident in the contrast between Marsh's utilitarianism and the eco-altruism reflected in the work of observers such as Thoreau sharpened in the closing decades of the nineteenth century, a development often typified by the influential figures of John Muir and Gifford Pinchot. Few writers argued for the preservation of the natural world more effectively than Muir, who based his appeal explicitly on the conviction that species exist for themselves and not simply for human use.[43] Espousing a kind of trans-species "Declaration of Independence," Muir was especially fond of sequoias, at least partly because they live for thousands of years and cannot be managed effectively. Yet he was far from envisioning wilderness as a "hermetic quarantine," and he hoped that people would visit wilderness, defining their contact with wild nature as a rejuvenating, perhaps even ritualistic or sacramental, experience.[44] He also acquiesced to some degree of human management of nature, praised the U.S. Army's administration of Yellowstone and Yosemite,[45] and jotted in his journal, shortly after spending several days in Yosemite with President Theodore Roosevelt in the spring of 1903, "Now ho! for righteous management."

What he meant by this is unclear, although environmental philosophers Bill Devall and George Sessions suggest that what he had in mind was "ecocentric ecological management," which they define as a nature-friendly approach to managing parks and forests, holistic in spirit and related to contemporary holistic forestry, which promotes reforestation, healing of damaged watersheds, and reintroduction of native species.[46]

Gifford Pinchot, on the other hand, opposed Muir's reverence for nature and trust in its self-healing ability with a managerial philosophy based on a utilitarian conception of the value of nature. As the first chief of the U.S. Forest Service, from 1905 to 1910, he argued that forestry is no more complicated than "tree farming—handling trees so that one crop follows another." Humankind has "a duty to control the earth he lives upon," Pinchot wrote.[47] Only years later did he concede that the value of a forest could not be measured solely in dollars and board feet.

Between them, these two men represent the two elements we find in tension with the idea of ecocentric restoration: on one hand a Franciscan respect for nature in its aspect as given and on the other a Benedictine willingness to manage it. In ecocentric restoration this would take the form of accepting the responsibility for managing land, but on its own terms, as defined by its historic condition and ecological trajectory independent of "us." These schools of thought resisted reconciliation or integration, however. In fact, they dominated the environmental thinking of the twentieth century and provided the two stools between which the idea of ecocentric restoration eventually fell.

Chapter 3

Preconditions

By the closing decades of the nineteenth century the basic elements were in place for the invention of ecocentric restoration. George Perkins Marsh had created and effectively articulated a story that cast people as separate from their environment to the extent that it made sense to say—or made it possible to see—that they could harm or degrade it. The formal closing of the American frontier in 1890 drew across history the line that eventually inspired and defined the idea of ecocentric restoration—the line between the nature we discover and the nature we inhabit and reshape. The romantics and transcendentalists had laid deep cultural foundations for the idea that nonhuman nature had value in its own right and so appealed not only to our aesthetic sense but also to our conscience. Landscape architects had brought the naturalizing school of design from its beginnings in the *bosco* of Italian Renaissance gardens and the naturalized landscapes and cottage gardens of designers such as Repton and Brown to the threshold that separates design from ecocentric restoration. And ecology was emerging as a science, represented by the end of the century by positions at a scattering of universities.

With this in mind, we have reached the time when we can start checking out projects, looking for what we might take to be the earliest examples of ecocentric restoration. Of course, origins and firsts are notoriously difficult to identify. And this is especially true for a practice such as ecocentric restoration, the invention and realization of which lie in part in the practitioner's intentions. Partly for this reason, the question of who invented this form of land management has no straightforward answer. Ecocentric restoration clearly developed piecemeal, in small steps, and with practice preceding realization of value and implications, in many cases by generations.

Our aim, then, is to explore these often small steps of invention, discovery, and realization, keeping in mind that we are not interested in restorative land management in general but only in the development of ecocentric restoration as distinct from these other forms of conservation.

This, we will see, is a story not of a great flash of genius—someone's "Eureka" moment—but of a series of small discoveries and realizations, more often than not by practitioners simply messing around with wildflowers or reflecting on the transformation of a landscape and the loss or degradation of a biome, practicing a version of gardening, trying to make it work, perhaps not even distinguishing it clearly from other forms of land management. It is a story not of a great watershed and wild surmise but of stepping-stones, of seat-of-the-pants experiments, modest insights, and small realizations, not by one or two but by dozens and even hundreds of people.

A Cliff-Face of Change

Both the practice and the idea of ecocentric restoration that began to take shape in the closing decades of the nineteenth century emerged from the experience of a particular culture responding to a distinctive environmental history. Europeans had come to what they thought of as a New World—Australia and the Americas—supposing it to be something like, if not literally, the lost Garden of Eden,[1] but they soon found themselves living on a continent that had undergone radical transformation in large areas as a direct result of their own enterprise.

This transformation was especially evident on the prairies of the midcontinent, where an entire biome had been transformed in the space of a single lifetime, beginning roughly in the 1820s. By the end of the century, millions of acres had been converted to farmland, and the biota that had composed what Walt Whitman had identified as "the emblematic heart of democratic America"[2] had been reduced to scattered remnants in back forties and along railroad rights of way. "Could he arise from beneath it," Willa Cather wrote of John Bergson, the Swedish settler of her 1913 novel *O Pioneers!* sixteen years after his death,

> he would not know the country under which he has been asleep. The shaggy coat of the prairie, which they lifted to make him a bed, has vanished forever. From the Norwegian graveyard one looks out over a vast checker-board, marked off in squares of wheat and corn; light and dark, dark and light.[3]

Of course, Cather's peaceful rural landscape, so recently "won" from the "wilderness" and its inhabitants, was a trope for the entire continent. At the beginning of the century Lewis and Clark had trekked from St. Louis to the Pacific as explorers, expected by Thomas Jefferson to bring back mammoth hides traded from the Indians, along with beaver skins. By its end, it was clear that there were no mammoths, even in the recesses of the great western mountains. Half a continent had been explored and settled by newcomers in the space of a single lifetime, and in the process vast areas of forest had been clearcut, wetlands drained, prairies, as Vachel Lindsay wrote, "swept away by wheat," and Huckleberry Finn's territory had become, as literary critic Leo Marx has said, Kansas City.[4]

Together with a parallel history in Australia, this was a unique event, not in its happening but in its speed and extent. Humans had transformed landscapes and even whole continents before, but usually slowly and with only limited means of keeping track of change over long stretches of space and time. "For many millennia," historian David Lowenthal writes, "most people lived under much the same circumstances as their forbears, were little aware of historical change, and scarcely differentiated past from present."[5] Even in Europe, where documents recorded changes in landscapes over periods of a millennium or more, these had been so gradual that the historic image they provided was a dim and incomplete picture of a series of gentle plateaus receding into the distant past. What large areas of the New World offered, in contrast, was unique in human experience: a historical cliff-face of unmistakable environmental change on a continental scale and within the short span of a human life. Joni Kinsey and her colleagues note that the midwestern prairies, which became something of a cradle and proving ground for the practice of ecocentric restoration a generation or two later, were settled and transformed roughly between the Black Hawk war of 1832 and the formal closing of the frontier in 1890, just fifty-eight years later.[6] A person born on the frontier, or "middle border," or brought to it as a child might easily recall the old prairies later in life, passing on the memory to children and grandchildren well into the twentieth century, as writers such as John Muir, Hamlin Garland, and Willa Cather did for a wide audience. And this transformation, not hinted at in dim records but lying within the horizon of living memory, whether celebrated as progress or deplored as desecration, naturally loomed large in the imaginations of the descendants of the pioneers and settlers who had brought it about.

Not surprisingly, after four centuries of "westering," this was an important psychological and cultural watershed. "With a considerable sense of shock," historian Roderick Nash writes in his pathbreaking 1967 book *Wilderness and the American Mind*, "Americans . . . realized that many of the forces which had shaped their national character were disappearing. Primary among these were the frontier and the frontier way of life."[7] One response, he writes, was a "wilderness cult" that changed the sign of wild nature from negative to positive. For a new generation, wild country was no longer just raw material, or an opponent or even an enemy to be overcome, but a valued other, indispensable as an arena for the incubation of the supposedly distinctive American virtues of enterprise, self-reliance, thrift, egalitarianism, and fair play. This added an active, red-blooded, populist, Teddy Roosevelt element to the romantics' discovery of nature as an inspiration and occasion for the experience of the sublime. But it also brought with it a sense of regret. In the space of a few decades, less time than now separates us from, say, the Korean War, midwesterners, repeating the experience of settlers from coast to coast in an especially dramatic way, found their habitat changed virtually beyond recognition. Nash writes that this experience of "contact and familiarity, followed by memory and nostalgia, created a new knowledge of (the prairie) and a new distance from which to view it," to the extent that *prairie*, which had meant something like *desert* just a few decades earlier, came to mean something like a lost Paradise garden. The mix of familiarity and emotional distance was crucial. As Nash points out, from the time of the first settlements in New England it was not the settlers themselves, busy with its transformation, who discovered the precontact landscape and recognized its enduring value but their children.[8] And this recognition was a response not to some newly realized utility or to nostalgia for a rural or pastoral comfort but to the memory, if not the experience, of nature not as "us" or "ours" but as wild and given and, as Kinsey and her coauthors write, "beautiful in its own right, with more than monetary value." This sense of value beyond the utilitarian was felt acutely by many of the conservationists of the period and was clearly part of the psychology underlying the idea of ecocentric restoration.

But if Americans, looking back a generation or two, felt nostalgia and regret for the loss of the wilderness settled and subdued by their grandparents and parents, they had considerable conceptual means for responding to this. One of these was what philosopher Bruce Wilshire characterizes as a distinctively American philosophy of nature grounded in the experience of the frontier and reflecting the intellectual influence of Native American peoples. Formally expressed in pragmatism, as developed by thinkers such as William James and John Dewey, this philosophy, Wilshire writes, effectively bypassed Descartes and modernism in philosophy to take the body, concrete reality, and nature seriously in a way that modern philosophy generally has not. Itself an expression of a more generally shared intimacy with nature-as-given fostered by the experience of settlement and the frontier, this formal philosophy reflected a state of mind open to the idea that other species have intrinsic value as subjects, to experiences that break down the distinction between subject and object, and to a conception of value that is ecological in the sense that it locates value in relationships rather than individuals.[9]

Another piece of conceptual software on the American hard drive that directly related to the development of ecocentric restoration was the habit of reading events in terms of the pattern of fall and redemption as represented in the biblical recovery narrative. Preached from pulpits, acted out in monastery gardens for centuries, and brought to the American colonies in the form of the myth of the restoration of first times we described in the previous chapter, this was a story about nature, humans, and their God that placed responsibility for both the Fall and the recovery squarely in human hands.[10]

This story, and the idea of nature, value, and goodness it expressed, provided the framework Americans used to interpret and make sense of wilderness and also of environmental problems such as those documented by Marsh and historic events such as the closing of the frontier. And if in the earlier version wild nature was bad, the story made it easy to turn this around. If, as Nash emphasizes, the Bible provided ample warrant for the idea of wilderness as inhospitable and even an abode of evil—the "howling wilderness" of the Puritans, a landscape needing redemption by human effort—it also, as more recent writers have shown, figured it as a place where humans experienced God in a special way, as a reflection of his goodness and even, as in the Exodus, a theater in which the drama of salvation would be played out.

This being the case, environmental redemption could mean either "reclaiming" the desert by refashioning it into the archetypal garden—essentially the opposite of ecocentric restoration—or it could mean restoring it by moving in the opposite direction, from the cultivated condition, now seen as disgraced, back to its original condition, fresh from the hand of the creator. This search for redemption through subtraction of civilization, figured as city, factory, or battlefield, has been an important theme of American literature, reflecting at a personal level the broad current of the American adventure itself, understood and experienced as an escape from history for the sake of "life, liberty, and the pursuit of happiness" in the bosom of nature.[11] And because it aimed at recovery of an imagined or idealized original condition, figured as Eden, it entailed not an exploration of history (a history that was actually crowded with politically, militarily, and ecologically potent indigenous peoples) but an escape from it. This was a fantasy, a projection of the imagination based on a delusion. But it was also a way of conceptualizing original nature as having original value—that is, value apart from its value as raw material for human exploitation. This notion is just a small step away from—indeed is a concrete and poetic articulation of—the more abstract idea that nature has inherent value and that we therefore have a moral obligation to preserve—and perhaps to restore—it.

In addition to this master recovery narrative, other factors, both internal and situational, were coming together toward the end of the century to favor the invention of a form of land management based on this way of valuing nature. These included distinctive elements of the liberal Protestantism that historian Donald Worster identifies as having played a key role in the shaping of American environmentalism, each of which is consistent with the idea of restoration in general and ecocentric restoration in particular.[12] Thus what Worster calls "moral activism" inclines people to take active steps to right wrongs once they are aware of them. "Ascetic discipline" predisposes them to look beyond immediate pleasures and advantages for the sake of the long-term good. "Egalitarian individualism" construes all people as equal in worth, an idea that, Worster notes, "once set in motion can prove exceedingly difficult to control" and "may lead not only to elevating the poor and despised in society, but also to investing whales, forests and even rivers with new dignity—to the discovery of the concept of the rights of nature" (198). This, of course, is consistent with the concern for "all the parts" that distinguishes ecocentric restoration from other forms of land management. And finally, "aesthetic spirituality." This is the idea that beauty lies not in the eye of the beholder but in living in harmony with nature, and it underlies a wide range of environmental values and practices, certainly including ecocentric restoration.[13]

Besides, these inherited ideas and attitudes, a number of developments in the decades around the turn of the twentieth century helped create favorable conditions for the development not only of conservation generally, with its commitments to various forms of meliorative restoration, but also of ecocentric restoration specifically.

One of the most important of these was a growing skepticism regarding the idea of progress and, complementing that, an increasing interest in the past during this period.

Historian David Lowenthal, who has done extensive research on Americans' relationship to their past, writes that Americans felt a general dissatisfaction with their condition between the end of the Civil War and the end of the nineteenth century. Lowenthal attributes this to anxiety over the accelerating pace of change; the emergence of ideas such as evolution and entropy, which challenged the idea of the regularity of nature; and growing dismay over "their land's lost purity and vanishing wilderness."[14] Confronting such changes, Americans began to realize that the old things are different; they belong to, as he puts it, "a foreign country."[15] And so they are valuable and of interest for that reason alone—not merely as resources or tools but because they are unique and also because, like wilderness, being different from things that are contemporary, familiar, and useful, they provide a way of defining who we are in the only possible way, in terms of what we are not or, in the dimension of time, what we were but no longer are.

The result was a new interest in preserving and restoring elements of the past that, Lowenthal writes, "transcended partisan purposes or personal nostalgia" (367) and so pointed away from meliorative and toward other-oriented, or altruistic, approaches to land management. For the first time there was a constituency for the preservation and restoration not only of elements of the past held to be useful, beautiful, or inspiring but also of the "whole story," including nature as well as history and including elements that are useless, unattractive, and even embarrassing. This eventually included slave quarters, industrial landscapes, and relics of the bombing of Hiroshima. It also included miasmatic, mosquito-breeding wetlands, habitat for dangerous predators, and ecological communities maintained by fire and composed mainly of "useless" grasses.

Origins

Looking around for inklings of the practice of ecocentric restoration in the closing decades of the nineteenth century, we find relevant developments in three overlapping areas: in various land management practices oriented toward resource conservation, in landscape architecture, and in the emerging science of ecology. We will explore these in this order, discerning in each of them three indistinct steps toward the practice of ecocentric restoration.

Resource Management

The resource management agencies that appeared in late-nineteenth and early twentieth-century America were largely the creations of the liberal reformers, usually identified as progressives, whose efforts reflected the utilitarian idea of conservation prevalent at the time.[16] The utilitarian philosophy that guided the development of the U.S. Forest Service under Gifford Pinchot was typical—indeed, conventional—and underlay conservation efforts by agencies such as the Soil Conservation Service, the Bureau of Land Management, and the Fish & Wildlife Service that took shape in the first half of the twentieth century.

An interesting exception, at least in principle, was the National Park Service, created in 1916. In contrast to the efficient use philosophy that guided the work of these other agencies, the national parks were set aside explicitly to preserve natural landscapes without regard to their utilitarian value, apart from their appeal to visitors. Yet for much of their history, the Park Service, emphasizing visitor satisfaction rather than conservation, managed the parks principally for the sake of visitors, a practice that park service historian Richard Sellars has called "façade management,"[18] and, as we will see, actually resisted the notion of ecocentric restoration until quite recently.

As far as the working landscape was concerned, conservation efforts came closest to ecocentric restoration in their aims and outcome when those responsible had—or were exploring—the idea that the historic system "worked better" in some practical sense than the existing, altered system, and so systematically added elements in various combinations in a deliberate search for the critical factors or combination of factors responsible. A good example is the work of managers such as Arthur Sampson and Lincoln Ellison, whose work on western rangelands Marcus Hall has discussed in some detail. Both Sampson and Ellison experimented with various forms of restoration in the belief—or hope—that certain historic systems in the arid West were better and more reliable suppliers of resources such as timber, forage, soil, and water than lands degraded by overgrazing. Responding to ranchers' calls for restoration of these earlier conditions, these researchers wound up interviewing early settlers and scouting out-of-the-way areas for clues to the composition of the "original"—and presumably stable—climax communities of the area. These attempts to restore natural systems were motivated by an interest in specific resources, however. Researchers chose them as models in the belief that they represented models of stability or sustainable productivity. This being the case, they were not ultimately committed to restoration of historically and ecologically authentic systems, and indeed they added exotic species to their plantings when they found it made them more successful, a practice they saw as in a sense more natural than nature itself because it accelerated the natural processes of species dispersal and community development.[19]

Perhaps. But of course this is not ecocentric restoration, which is not about creating natural capital but about recreating the old system. This may, and often does, entail actual sacrifices of resource value and stability, and resource conservation, important as it is, will not lead there. To find the currents leading in that direction, we have to turn elsewhere, specifically to the naturalizing school of design dating to the late eighteenth century and to the science of ecology. Indeed, it was the coming together of these disciplines in the decades around the turn of the twentieth century that resulted in the "invention" of ecocentric restoration. What this entailed was, first, a commitment to the representation of natural landscapes by designers, then the rejection of design and the replacement of aesthetic motives by the more abstract criteria of ecology as a basis for defining objectives. Significantly, both developments took place neither in rural or working landscapes nor in remote wilderness areas but in the cultural and intellectual playgrounds represented first by intensively used parks, usually in urban and suburban settings, and second by outdoor "laboratories" created by ecologists and nature aficionados with a gardening bent.

Landscape Architects

Reflecting the turn toward nature around the beginning of the twentieth century, cities began placing more emphasis on the creation of parks, and designers began incorporating native plants and natural vegetation in their projects as a way of bringing nature into cities.

The iconic example is New York City's Central Park, designed by Frederick Law Olmsted and Calvert Vaux in the years after 1857. This brought an interpretation in the manner of the English natural landscaping tradition into the heart of the nation's largest city. Even more interesting for us, however, was the project the two designers outlined in 1887 for the Niagara Reserve around Niagara Falls, which just two years earlier had been established as the first state park in the United States. The plan is of special interest for us because it was aimed at conservation of natural resources, not in the usual sense but only in the sense, which Olmsted and Vaux expressed very clearly, that the primal landscape—even when reconstructed—had value for visitors simply for what it was, even if it turned out to be in some ways inconsistent with popular taste. The project is also of special interest to us because it provides an opportunity to examine exactly how the idea of re-creating a natural landscape was conceived and how it was carried out at a time when the science of ecology barely existed and had nothing to contribute to an enterprise of this kind.

Except for this crucial missing element, the Niagara site offered conditions congenial to the invention of the idea of ecocentric restoration. The falls were a spectacular natural feature that was widely valued *as* a natural feature. And both the falls and the surrounding area had been seriously altered—"disfigured," as French geographer Élisée Reclus put it in 1874— by "hideous buildings, mills, workshops, hotels and warehouses . . . a depravity of taste."[20]

In fact, Olmsted and Vaux's plan reads in many ways like a manifesto for something like ecocentric restoration. Most dramatically, it called for the removal of some 150 buildings and structures, insisted on the use of native plants set out in naturalistic groupings, and asserted that exotic species that might "add to the interest of the public" should not be used to replace native vegetation damaged by landslides, ice, or insects, because these would be "as undesirable as the ornaments of stained glass, cut stone, plaster, paint and fountains that your Board has been removing." Anticipating similar language in the definition of restoration promulgated by the Society for Ecological Restoration more than a century later, they asserted that the resulting system would be largely self-sustaining, suggesting that "when all that is proposed is fairly done there will be no need for any fresh appropriations for construction. The work henceforth will be, strictly, a work of maintenance." They argued strenuously for the value of the historic landscape that they aimed to evoke and expressed concern that the project would always be vulnerable to schemes for "improvements" along lines that reflect a failure to distinguish between the aim of restoring the natural scenery and those of "ordinary gardening works," which they pointed out are designed to "give pleasure in a very different way."[21]

In all these ways, Olmsted and Vaux's plan for Niagara Reserve seems consistent with the idea of ecocentric restoration as we have defined it. Yet it was not quite an ecocentric restoration project, either as conceived or as carried out, because, though in some ways challenging conventional taste, it was undertaken for the sake of visitors and not for the sake of the historic landscape. (Commentators sometimes blur the distinction, supposing that, for example, aesthetic objectives are not utilitarian and self-interested. But of course they are,[22] and the point is an important one, as we will see when we consider the fate of restoration in the national parks.) Indeed, considering the Niagara project in relation to the idea of ecocentric restoration, Olmsted scholar Charles Beveridge points out that the predominant consideration reflected in Olmsted and Vaux's plan was not restoration or reproduction of a landscape at all but rather creation of "a parklike setting for viewing a remarkable natural feature". In fact, Beveridge notes that the project came closest to ecocentric restoration not on the mainland area most often described as restored but on nearby Goat Island and the Sister Islands, where the designers "sought to recover (or at least approximate) the special richness and density of the (historic) vegetation." Overall, Beveridge characterizes Olmsted and Vaux's work at Niagara as "creation of spaces drawing from the character of natural landscapes of certain kinds that had a powerful therapeutic and restorative effect on their users." Overall, he comments, Olmsted in particular "seems to have agreed with Thoreau in not caring for wilderness, but in seeking wildness"—that is, not a particular ecosystem but a certain quality associated with it.[23]

Olmsted and Vaux's projects at showcase sites such as Central Park and Niagara Falls proved to be the influential leading edge of a school of landscape design that rapidly developed in the decades that followed into a school of landscape design that received popular elaboration in Frank A. Waugh's 1917 book *The Natural Style in Landscape Gardening*. A leading academic authority on landscape architecture in the early twentieth century, Waugh was inspired by the "spirit of the natural landscape." The natural style, he wrote, sought to "present its pictures in forms typical of the natural landscape and made vital by the landscape spirit."[24] Over the years Waugh refined his ideas about the natural style, emphasizing the importance of topographic forms and arrangement of plants in masses imitating the patterns found in natural vegetation. In the Midwest, especially in the Chicago area, landscape designers and architects of the Prairie School used the city's parks to forge a connection with the region's historic landscape. Members of this group, notably Jens Jensen and Ossian Cole Simmonds, pioneered this movement, and their work undoubtedly influenced Waugh's writings on the natural spirit in landscape design.

The origins of the prairie school were humble enough. In 1888, Jens Jensen, a recent Danish immigrant and landscape designer at Chicago's Union Park, planted his "American garden" of perennial wildflowers set against a backdrop of native trees and shrubs, a practice he may have picked up from Olmsted, H. W. S. Cleveland, and others who were introducing naturalistic forms in their work.[25] Over the next several decades he celebrated the midwestern landscape in his work, seeking to awaken people to the beauty and diversity of a natural landscape that was rapidly disappearing as cities such as Chicago expanded.[26]

Jensen's parks were really natural gardens, symbolic of larger landscapes. When he redesigned Chicago's Humboldt Park beginning in 1896, he took an important step in the direction of ecocentric restoration by including a new element, the prairie river, and whole communities of wetland plants. At Lincoln Memorial Garden in Springfield, launched in 1936, Jensen provided an example of an explicitly historical motif by creating a landscape "such as Abraham Lincoln himself might have seen." In 1900, Jensen had the good fortune to meet Henry Chandler Cowles, a recent University of Chicago Ph.D., trained as a botanist and plant ecologist, and their close association contributed immensely to Jensen's knowledge about native plant communities.[27] The design at the Springfield site reflects this knowledge and is especially remarkable because it took into account the crucial element of ecological dynamics. Previously, landscape architects had paid little attention to dynamics and had created plantings intended to represent a finished condition. In his design for the Springfield project, however, Jensen, aiming to imitate the dynamics as well as the composition and structure of the vegetation types he was using as models, called for plantings that would serve as "a framework for triggering a series of successional changes in the landscape that have resulted in the landscape mosaic that exists today."[28] Moreover, because the project relied heavily on volunteer labor, it was also a forerunner of the volunteer-oriented restoration efforts that gained prominence decades later.

Ossian Cole Simonds shared with Jensen an appreciation of the prairie landscape,[29] and in addition to his work at Chicago's Graceland Cemetery and Morton Arboretum, he experimented with restoration on land he owned on Pier Cove on the eastern shore of Lake Michigan, undertaking re-creation of a beech–maple woods and a number of experimental pine plantings.[30]

In his 1915 book *The Prairie Spirit in Landscape Gardening*, Wilhelm Miller recognized the "prairie movement" as a distinctive American style of landscape architecture that "sought to re-create as much of the local scenery or vegetation as practical." And he pointed toward what we would call the performative dimension of this work by suggesting that the scale of a project, which was usually small, is perhaps incidental to the restorer, who demonstrates through this gesture that "he wants to be surrounded by common, native things, rather than by rare and costly foreigners."[31]

Ecology

Although Miller, Jensen, and Simonds shared a commitment to the restoration of regional landscapes, their work provides only a partial precedent for ecocentric restoration. What we see in projects like theirs is the naturalizing school of landscape design and wildflower gardening approaching the ideal of ecocentric restoration, even at times reaching it, as Jensen arguably did at Lincoln Memorial Gardens. In most if not all of these projects, however, ecological and historical accuracy was subordinate to human interests. Ecocentric restoration could not happen until the order of priorities was decisively reversed and design considerations explicitly set aside in favor of a commitment to literal reproduction—not of landscape as interpreted, however sensitively, by a designer but as an ecological community or ecosystem described, however imperfectly, by the scientist and the historian. Both disciplines reflect the objectifying principle to which a science such as ecology aspires, and together they brought to the art of landscape design the commitment to impartial and inclusive description and representation that distinguishes copying from imitation, reproduction from improvisation, and ecocentric from more creative and self-interested forms of land management. In a very real sense, they took away the art, and this did not happen easily. In fact, as we will see with special clarity as these competing visions encountered each other in the national parks, they led to tension and outright conflict between professions.

An important step in the direction of restoration of whole ecosystems was the creation, around the turn of the twentieth century, of what amounted to living dioramas, plantings representing, with varying degrees of completeness, specific plant communities, not (or not exclusively) as amenity landscapes but as examples of a vegetation type. Notable examples were projects at the Garden in the Woods in Massachusetts, the Brooklyn Botanic Garden in New York City, the Strybing Arboretum and University of California Botanic Garden in the San Francisco Bay area, the Geographical Arboretum of Tervuren in Belgium, Kirstenbosch National Botanical Gardens in South Africa, and the Meiji Shrine in Tokyo.

However, these typically focused on specific elements of the plant community, in many cases trees. They did not include animals and, conceived in geographic rather than historic terms, were not really restorations but creations. Though planted in naturalistic associations, they were generally conceived more as a collection of plants than as comprehensive representations of actual ecosystems. It was ecologists who, coming online in this period, brought in the emphasis on the whole ecological system as characterized not by an artist, resource manager, or botanist but by an ecologist.

Ökologie—the word was coined by German zoologist Ernst Haeckel in 1866—developed from little more than a label to a distinct and productive branch of biology in the United States in the decades we are considering here. This was roughly from 1893, when the term was formally adopted by a group of botanists meeting in Madison, Wisconsin to refer to the study of the physiology of plants in their natural habitats,[32] into the 1920s and 1930s, when ecologists played key roles in a number of restoration projects that collectively constitute a watershed in the invention of ecocentric restoration.

At the same time, in addition to its formal commitments to the scientific investigation of relationships between organisms, ecology, as it developed in the United States during this period, had a number of characteristics that were especially relevant to the development of what Aldo Leopold later called "a science of land health" in general, and the practice of ecocentric restoration in particular. Together, these strongly influenced restoration, contributing to its development but in some ways limiting the full realization of its distinctive value.

Eugene Cittadino, a historian of ecology, points out that from the beginning ecologists emphasized the relevance of their discipline to practical matters ranging from agriculture to forestry, fisheries, and game management. Underlying this utilitarian commitment, however, was an idea of creation that reflected Herbert Spencer's idea of evolution as progress toward an ever more stable and harmonious association of organisms. It also included an ethic of stewardship deeply rooted in a Protestant version of the idea that nature is a book that reveals the nature of God and so has value for its own sake, apart from human or utilitarian considerations.[33] Historian Mark Stoll notes that half a century before Leopold wrote of land as a "community to which we belong," these ecologists, together with the social scientists whose discipline was taking shape at the same time and with whom they exchanged ideas, "believed firmly in the intrinsic *value* of the subjects they studied (the natural world and society, respectively)—as opposed to studying them solely for their economic or social utility."[34] In other words, these pioneers thought of the biomes they investigated, whether a lake in Illinois or a prairie in Nebraska, as inherently valuable communities of creatures to which humans had a moral obligation. This essentially religious idea of the value of the subjects of ecological investigation complemented their scientific commitment to develop comprehensive and accurate descriptions of biomes. Taken together, these rules of engagement, so to speak, provided both the conceptual and the moral basis for, first, the preservation and then the restoration of ecosystems that would be ecologically complete and historically accurate, including all their elements, whether these were regarded as having value for humans or not.

A second characteristic of American ecology, also reflecting its roots in Protestant theology, was the idea of the purity and innocence of nature set off against the inherent sinfulness of fallen humans, who inevitably disturb the natural order but are responsible for correcting these disturbances.[35] This notion of an ideal original condition of creation lost and its restoration as the key to redemption—the ecologist's version of Hughes and Allen's "myth of the restoration of first times"—has important implications both for ecology and for restoration. In the hands—or rather the heads—of ecologists, this took the form of the idea that nature, uncompromised by humans, moves naturally toward order and stability. "There is a general consent," ecologist Stephen Forbes wrote in 1880, echoing Marsh, "that primeval nature, as in the uninhabited forest or the untilled plain, presents a settled harmony of interaction among organic groups which is in strong contrast with the many serious maladjustments of plants and animals found in countries occupied by man."[36]

This idea of the integrity of the natural or original association that could only be harmed by the influence of fallen humans influenced ecology profoundly, most notably in Frederic Clements's idea that the ecological community is an organism-like entity that, if disturbed, moves autonomously back toward a stable, climax condition. This implied that such associations were not merely old or historic but actually existed outside historic time, in the cyclic time implicit in the idea of redemption through the restoration of first times, itself a version of what comparative religionist Mircea Eliade called the myth of the eternal return.[37] This played an important, if ultimately ambiguous, role in the development of ecocentric restoration. For one thing, it encouraged the idea that all the parts of an association are not only essential to its proper functioning, but also *belong* there in some more fundamental, moral, cosmogonic, or religious sense. For another, it made the old, climax association an attractive model for restoration efforts, not only because it gave it an ontological status denied it by more individualistic theories of community dynamics but also because its self-organizing capacity would presumably aid and abet the restoration effort, pulling the system toward the ecologically privileged condition represented by the model system.

At the same time, by attributing desirable qualities such as beauty and stability to these original associations, it implied that restoring them would, almost by definition, benefit humans. This favored the idea of restoration, but because it rested on the assumption that, once restored, these biomes would be an ideal, even edenic human habitat, useful, productive, stable, attractive, and pleasant to be in, it actually stood in the way of ecocentric restoration because it overlooked or downplayed the idea that land management for the sake of the old ecosystem might entail economic, aesthetic, or emotional sacrifices as well as benefits for humans. Indeed, one reading of the biblical creation story held that all of nature—not just humans—had been disgraced by the Fall.[38] This supported the idea that landscapes such as deserts or mountains that people happened to regard as useless or unattractive are actually a maimed nature, victims of the Fall, awaiting redemption by human ministration. Understood in this way, reclamation of such landscapes meant the diametric opposite of what we mean by ecocentric restoration—meant, in fact, its destruction by, say, irrigation and cultivation intended to restore it not to a historic but to an ideal—indeed mythic—condition. In a milder form, this was the idea that would inform thinking about the national parks as "pleasure grounds," an idea that precluded a policy of ecocentric restoration in the parks for the better part of a century. It is also an idea, deeply inscribed in American thinking, that has made it difficult for environmentalists to commit themselves to, or even to recognize, the value of ecocentric restoration as a tribute to the inherent value of biomes apart from human interests, even after the practice of restoration, if not the idea of ecocentric restoration, has been accepted, articulated, and espoused by several generations of environmentalists.

But even if ecology, like any human enterprise, came into being and went about its business imbued with often unrecognized assumptions and mythologies, it was ecology, coming of age in the first few decades of the past century, that provided the conceptual software needed to make the move from versions of meliorative restoration that increasingly resembled ecocentric restoration to ecocentric restoration itself. Although they would never fully succeed, the ecologists would at least play the game of stepping back, trying to see past or through mythologies and other preconceptions in order to inventory what makes up an ecosystem and describe what is going on in it. The notion of attempting to reproduce such a system is a logical next step—a fascinating, indeed seductive, way of reducing to practice and trying out the ideas emerging from this new way of looking at nature. We find ecologists doing this in two contexts in the early decades of the twentieth century. Some took on the task of maintaining existing ecosystems, such as those in the national parks, in their original condition, a task that they began to realize entailed active attempts to compensate for novel influences on those ecosystems and so amounted to restoration. Others attempted to re-create historic ecosystems wholesale, on a much smaller scale, for explicitly scientific or educational purposes and on sites drastically altered by activities such as logging or farming. Both played important, complementary roles in the invention, discovery, and realization of the idea of ecocentric restoration.

The National Parks

In retrospect, the national parks offered the most likely opportunities for the development of ecocentric restoration because the legislation that created the National Park Service, the Organic Act of 1916, stated that the purpose of the parks would be "to conserve the scenery and the natural and historic objects and the wild life therein and to provide for the enjoyment of the same in such manner and by such means as will leave them unimpaired for the enjoyment of future generations."[39] Because ecologists and land managers eventually realized that leaving an ecosystem "unimpaired" would entail active management to maintain it in its historic condition or on its historic trajectory, this implicitly defined the national parks as the kinds of places where ecocentric restoration eventually would be practiced. And in fact, as soon as ecologists showed up they began pressing for what amounted to a program of ecocentric restoration for the parks. Ecologists were not part of the picture in the national parks for roughly half a century after the creation of the first parks and a decade after the passage of the Organic Act, however. And when they did enter the picture, their recommendations ran into a wall of institutional resistance that blocked their systematic implementation for another half century. Altogether, the story of how ecocentric restoration fared in the national parks provides perhaps the clearest example we have of the fate of this idea and the forces both promoting and retarding its development over a period of seven decades—a historical transect, as it were, cutting across three generations of conservationists and at least two major environmental movements.

What we find here is a story of an idea taking shape and struggling for realization in a real-world context of evolving and competing ideas of nature, complicated by institutional inertia and turf battles between professions and schools of thought.

Of these obstacles to realization, the most fundamental was ultimately a matter of definition. Although the Organic Act required that the parks be left "unimpaired," meaning, as Richard Sellars notes, essentially in their natural condition, what this meant in practical, operational terms was far from clear. *Natural* is a notoriously ambiguous term, and in the absence of less ambiguous ecological and historical criteria for defining management objectives, park managers generally regarded as "unimpaired" any area where existing ecosystems had not been destroyed outright or altered in conspicuous ways. Americans might have created the parks to preserve nature, as Marcus Hall has argued. But what that meant was hopelessly unclear. "There seems," Sellars writes, "to have been no serious attempt to define what it meant to maintain natural conditions." As a result, "This key mandate for national park management began (and long remained) an ambiguous concept related to protecting natural scenery and the more desirable flora and fauna."[40]

Although popular mythology has it that the idea for the national parks was concocted in "a moment of high altruism" by a group of campers inspired by the grandeur of the scenery near the origin of the Madison River in present-day Yellowstone National Park in the fall of 1870, the actual history is more complicated and included a large element of commercial interests. Not least of these was the interest of the Northern Pacific Railroad in developing tourism in "underused" wilderness areas in the West. These two visions for the park were similar enough to create an effective constituency—effective enough to lead Congress to create Yellowstone, the world's first national park, just a year and a half later. But they were by no means congruent ideas and, together, provided the conceptual foundations for an initiative, an agency, and a mission built precisely on the fault line between nature and culture and between the old self-interested idea and the altruistic idea of nature from which the idea of ecocentric restoration eventually emerged.

What prevailed in the circumstances, shaping National Park Service policy from coast to coast, was an idea of nature that was basically subjective, romantic, sentimental, and utilitarian and that underlay a program to manage the parks for the sake of their human constituents. This included the outright exploitation—and management—of resources such as water, timber, fish, and even minerals. But it also included management of parks as amenities, encouraging tourism by managing resources and viewscapes in accord with the prelapsarian, postcard idea of wild nature that was conventional at the time.

From the beginning the aim of the parks was famously schizophrenic—to preserve for use—and this left the door open for a wide variety of conflicting interpretations. Even management for "passive" use of parks entailed neglect of most elements, together with highly selective manipu-

lation, including expulsion of indigenous peoples,[41] a perverse move ecologically as well as morally because, as is now widely recognized, pre-
Columbian peoples had played an important role in shaping many of the
ecosystems places such as the national parks were intended to preserve.[42]
It also included practices such as predator control, stocking of fish (including nonnative species), fire suppression, insect control, and winter
feeding to favor popular species such as bison and elk.[43] This bore little re-
semblance to the kind of disinterested management that would be necessary to maintain the parks on their "natural" or "original" (more accurately, historic) trajectory. This was as true of the attempts to enhance the
scenery as of the efforts to tinker with wildlife populations in order to bring
parks into line with what visitors expected to find there. Marcus Hall inter-
prets landscaping efforts that favored native plants and naturalistic interpretations in some park areas as a step toward the development of a
restoration-oriented approach to management of park ecosystems.[44] But
this is a historical elision that seriously misrepresents how the Park Service
actually responded to the prospect of ecocentric restoration. Where Hall
depicts continuity, there was actually protracted discontinuity reflecting
conflict between two cultures committed to radically different ideas of
value in nature.

Far from representing a step toward a policy of ecocentric restoration,
both the landscaping and the management of ecosystems done in the
parks during this period were at odds with the idea of ecocentric restoration in ways that had important ecological consequences for the parks.
Both were human-centered rather than ecosystem-centered. The landscaping carried out in high-use areas such as roadsides, campgrounds, and
popular viewscapes, amounting to Sellars's "façade management," reflected popular taste and had little to do with the management of actual
park ecosystems. And the management efforts that were carried out on behalf of park ecosystems focused on popular species such as elk, bison, or
bears, often at the expense of predators, creating the legacy of extirpations
and irruptions that attracted the attention of conservationists such as Aldo
Leopold and George Wright in the 1930s and that would remain an issue for the parks throughout the century. Both were the practical expression of a philosophy of park use and management that dominated the Park
Service for much of its history and was directly opposed to the other-
regarding, science-based management that would constitute a program
of ecocentric restoration. As Sellars comments, scientists played almost
no part in the management of the parks before the 1920s and only a limited and contested one for decades after that. During that period, land

management policies were shaped by administrators and landscape architects who were concerned primarily with visitor satisfaction and "gave no substantive consideration to an exacting biological preservation" of park ecosystems. Frederick Law Olmsted Jr., the son of the Olmsted of the Central Park and Niagara Reserve projects, who did consulting work for the parks during this period, "rarely even alluded to preserving natural conditions," Sellars writes. "He seems never to have seriously considered the parks as having anything like a mandate for truly pristine preservation" (45).

It was, necessarily, scientists who challenged this subjective approach to ecosystem management in the parks. And it was the Ecological Society of America, then just seven years old, that drew early attention to the ecological risks and damage associated with the introduction of nonnative species in the parks by passing in 1921 a resolution urging that such introductions be "strictly forbidden." The American Association for the Advancement of Science passed similar resolutions the same year and in 1926. The first serious challenge to park policy from within the Park Service came almost immediately, in 1928, when the agency for the first time—and only, Sellars notes, with the promise of private funding—hired a handful of scientists to address management questions (87). The key figure in the development of what Sellars calls a "truly revolutionary" (148) vision for the management of park ecosystems was a young biologist named George Wright.

Wright, then twenty-four and an assistant naturalist at Yosemite National Park, proposed—and, being independently wealthy, offered to fund—what over the next decade and a half developed into the first systematic survey of wildlife in the national parks. The result of this initial field work, begun in May 1930, was a 157-page booklet published in 1933 and titled *Fauna of the National Parks: A Preliminary Survey of Faunal Relations in National Parks*. Familiarly known as *Fauna No. 1*, the report proved to be what Sellars calls "a landmark document" that proposed "a truly radical departure from earlier practices" in calling for what was, in effect, a comprehensive program of ecocentric restoration for the parks. Specifically, the biologists began with the mandate that the parks be maintained in their natural condition, but for the first time they defined *natural condition* in ecological and historical rather than aesthetic terms. From this perspective, park ecosystems that had seemed unimpaired when viewed simply as natural areas, turned out, when scientists actually counted elk, bison, wolves, bears, cougars, and coyotes, to be seriously impaired, in the sense that they differed dramatically from their historic condition. Looked at in this way, the parks were a mess. Indeed,

Wright and his colleagues saw what they described as "a very wide range of maladjustments" in populations of animals ranging from elk to pelicans. In fact, *Fauna No. 1* coauthor Ben Thompson declared in a memo to director Arno Cammerer early in 1934 that "no first or second class nature sanctuaries are to be found in any of our national parks under their present condition."[45]

It is this recognition of alteration, revealed by an actual inventory of components and processes, that proved to be the essential first step toward the idea of ecocentric restoration. Decrying the policy of scattershot, selective, and hands-off management, Wright and his colleagues urged management based on research and, "proposed . . . where necessary and feasible, to *restore* park fauna to a 'pristine state'" (96). Clearly reflecting the scientist's distanced perspective on nature as an object of study, if not a subject having value in its own right, they insisted on management of the biome as a whole, including all its elements. This would include not only predators such as coyotes and bears, which agency managers had often targeted for reduction or elimination, but also nonmammalian, noncharismatic species such as rattlesnakes, which posed some danger to visitors.

As these stories illustrate, the commitment not only to restore a biome but to define the objectives of restoration in ecological rather than aesthetic or economic terms gave a hard edge to thinking about ends and means and immediately raised questions that those considering restoration and management from a utilitarian and aesthetic perspective answered in different ways, or had not asked at all. Eschewing both economic and aesthetic criteria for defining and evaluating the condition of a biome, and anticipating questions restorationists would debate three quarters of a century later, Wright and his colleagues confronted the question of how to choose historic models, acknowledging that because biomes are always changing, there is "no one wild-life picture which can be called the original one."[46] Their answer was the one that characterized much of the ecocentric restoration work carried out in New World settings in subsequent decades: They noted that so little was or could be known about the pre-contact era, and that so much had changed since that time, that "the situation which obtained on the arrival of the (European) settlers may well be considered as representing the original or primitive condition that it is desired to maintain."[47] This condition at the time of cultural contact, the distinctive New World experience, became the favored objective of restorationists in later decades.

Wright and his colleagues also confronted the question of human interference in a natural biome, urging minimal interference but also arguing that the admission that "there are wild-life problems is admission that unnatural, man-made conditions exist. Therefore, there can be no logical objection to further interference by man to correct those conditions and restore the natural state."[48] They noted that, with the possible exception of actual islands, no park is an *ecological* island; all are subject to external influences, and none is large enough to support self-sustaining populations of "all resident species."[49] And, following up on the *Fauna No. 1* study at a superintendents' conference in 1934, Wright even acknowledged the impossibility of the task the report set for the agency, noting that it would be impossible to keep "any area of the United States in an absolutely primeval condition" but that "there are reasonable aspects to it, and reasonable objectives that (the Park Service) could strive for."[50]

As for actual management to maintain these "whole biotic superorganisms," the biologists recommended exactly the mix of letting alone and compensatory management that characterizes ecocentric restoration. They recommended that every species be left to "carry on its struggle for existence unaided" unless—striking the restoration note—they are threatened with extinction (97–98). In practical terms, this would mean ending the animal-feeding programs that park managers carried out to support populations of popular species such as bison, except in emergencies. It would also include the two key prongs of ecocentric restoration: elimination of exotic elements and reintroduction of those that had been extirpated from an area. This included species, of course, but also fire, at least to the extent that the biologists opposed the policy of clearing downed timber to reduce fire frequencies (128ff).

As for the human side of the work, it is interesting to see how those associated with this project articulated their motives with respect to the ecosystems they were working with. The overall impression is of a group of people arguing for restoration for the sake of the ecosystem but not quite saying so, no doubt partly because the idea challenged deeply engrained agency policy but perhaps also because the biologists hadn't really formulated this idea, even in their own minds. Sellars notes that in *Fauna No. 1* Wright and his colleagues "remained loyal to traditional attitudes, stating that public use 'transcends all other considerations,'" but adds that, looking ahead, they argued that the "most farsighted" policy would be "to minimize the disturbance of the biota as much as possible" (97). They also expressed their concern that the parks "not supply mass outdoor recreation," which they felt would place "a destructive burden" on the parks and that "giving all of the people everything they want within the parks . . . would involve sacrificing the Service's highest ideals" (139). These might call for some inconvenience—and even sacrifice—on the part of the public, as

Wright and Thompson implied when they noted with a bit of irony, in *Fauna No. 1*, that "it is easier to make the human adjustment to new circumstances than to coerce the animals."[51]

Overall, the picture that emerges is one of ambivalence. Wright and his colleagues were dealing with what they identified at the beginning of *Fauna No. 2* as "the whole difficult problem" of finding a form of land use that "permits neither the impairment of primitive wildlife nor the restriction of human occupancy."[52] This is a problem that managers have struggled with ever since, and it would eventually find at least a partial solution in the discovery of the value of restoration itself as a public activity.

Sellars writes that *Fauna No. 1* soon became the bible for the small contingent of park biologists, an endorsement that director Arno Cammerer made official by confirming its recommendations as agency policy in March 1934. However, this was by no means the happy ending of the story. The support for the ideas outlined in the *Fauna* series lasted only briefly. Wright himself died in an auto accident in New Mexico early in 1936, and, deprived of their most effective champion, the biologists quickly lost traction within the agency. New kids on the block in any case, they remained a tiny minority of Park Service staff, and lacking a public constituency favoring their ecosystem-oriented management philosophy, they had little bargaining power in debates about management policy. The visitor-oriented programs, including education, recreation, and landscape management, resisted from a position that had dominated agency policy for more than half a century.

Ultimately this first round in the contest between ecocentric restoration and a more general, vaguely defined program of ecosystem management came to an end in 1940, when Park Service biologists, then numbering only ten, only four of whom were funded from regular Park Service appropriations, were transferred out of the Park Service and into the Bureau of Biological Survey. It was a case of an organization expelling what it perceived as a foreign body. The problem was at bottom a difference in basic ideas about nature and its proper use embodied in different professions, which resulted in an oil-and-water relationship between them. Shifting from an essentially human-centered management philosophy to an ecocentric philosophy reflected in ecologically defined objectives would have meant challenging not only agency policy but also a popular aesthetic conditioned by a century of romantic depictions of landscape and wilderness experience.[53] It would also have entailed a major reordering of the social structure of the Park Service itself, with legions of rangers and educators committed to serving a public perceived as customers and backed by three generations of policy and tradition deferring to a handful of scientists with very different ideas about how to manage the parks.

This was not going to happen easily and, as we will see, did not happen for about half a century. What this reveals is the profound disjunct between the ancient idea of self-interested land management and the new idea of ecocentric restoration, which was just beginning to take shape in the minds of a few managers and biologists. As early as 1925 biologist Charles Adams had recognized that while forestry flourished in the national parks, in part because it had the advantage of a strong precedent in Europe, the national parks represented "a distinctly American idea" and that "these wild parks called for a new profession, far removed indeed from that of the training needed for the formal city park or that of the conventional training of the forester."[54] Summing up the situation after a decade and a half of effort by the biologists, whom he describes as "insurgents in a tradition-bound realm," Sellars writes, "Unlike the perspectives of the landscape architects or foresters, the wildlife biologists' vision of national park management was truly revolutionary, penetrating beyond the scenic façades of the parks to comprehend the significance of the complex natural world." Realization of this vision would come in other contexts, however, and would take many years.

Chapter 4

Invention

Ecocentric restoration, understood as the attempt to compensate for novel influences on an ecosystem in order to allow the system to continue on or to resume its original, or natural, trajectory, was clearly conceived by biologists such as George Wright and his colleagues. But it is hardly surprising that it failed to take root as a viable idea or accepted practice in the context of a large agency such as the National Park Service, where a well-established institutional culture oriented toward the satisfaction of public tastes and interests sturdily—and effectively—resisted what it rightly saw as a challenge to its basic assumptions about the agency's mission. However, during the same period that Wright's group and a few other like-minded biologists were pressing for a program of ecocentric restoration in the national parks, a handful of other scientists, experimenting with the same or similar ideas under very different conditions, actually managed to realize them in a practical way. They did this not on the grand scale of a national park such as Yosemite or Shenandoah but on the small scale appropriate for a new, and in many ways impractical, form of land management. And they did it either, as Ossian Simmonds had done, on privately owned land or under conditions in which, whether benefiting from the freedom of the private institution or the ivory tower privilege of the academy, they were as free as kids playing in a sandbox to tinker and experiment on their own terms, even when this seemed nothing more than fooling around or chasing pie-in-the-sky daydreams.

Because, of course, the daydreams were as important as any amount of practical expertise, if not more so. It is worth noting that most of the pioneering attempts at ecocentric restoration that we have identified were carried out in what we might characterize as a spirit of play, in many cases by amateurs or by professionals working at the margins of their disciplines and in a context protected from the insistent pressures of markets, politics, and constituencies. Indeed, this avocational element seems to have been crucial in allowing a scattering of land managers to cross the boundary between meliorative and ecocentric restoration. Ecology—science—was a crucial factor. But science itself, as Jacques Barzun has pointed out, is at bottom a "glorious entertainment."[1] Australian Ambrose Crawford, for example, was a dairy farmer who learned the plants he was working with by sending specimens to professional botanists. Harvey Stork and D. Blake Stewart at the Cowling Arboretum in Minnesota smuggled in their tiny forest restoration project as a footnote, apparently barely accounted for, in the context of a much larger project. Aldo Leopold tinkered with restoration on weekends at his family's rural retreat, and even his work with his colleagues at the UW–Madison Arboretum entailed at least to some extent a respite from "practical" considerations.

Ecocentric restoration, in contrast to meliorative land management, often *is* impractical; or at least includes an impractical element. In a sense it had to come out of a sandbox, a skunkworks or back 40. Looked at from the perspective of this small-scale, in many ways privileged tinkering, the futility of the efforts in the national parks is obvious. It was as though the Wright Brothers had attempted to make their first flight from the deck of an aircraft carrier in a high wind with a large audience of stockholders. For a first flight of any new idea, something much more modest—and much more playful—is called for. Besides this, given the ways in which the notion of ecocentric restoration challenged conventional ideas of conservation, aesthetics, the value of nature, and human relations with the rest of nature generally, it also had to be discreet. And so these early projects were carried out, like the Wright Brothers' earliest flights, far from the push and shove of agency politics and public accountability.

Partly for this reason, locating original projects is difficult. Quite understandably, those involved in these projects seem to have regarded them as demonstration projects rather than as experiments in a distinctive new form of land management that might have important implications for conservation. None were widely publicized or attracted much attention at the time. And most were short-lived. Of the six we have identified as early projects, started between 1906 and the mid-1930s, only one—at the UW–Madison Arboretum—has continued any kind of intensive restoration effort down to the present. This being the case, the handful we have identified as starting up during this period presumably represent only a sample of those undertaken, as fish or birds counted in a census represent only a fraction of those actually present. However, these projects defined their objectives in a way that was distinctive in much the same way that George Wright's prescriptions for Yosemite were distinctive—that is, in their commitment to privilege ecological over aesthetic and other human-centered considerations in defining objectives. Small as this sample may be, what it suggests is that the conservation community was reaching a tipping point during this period, internalizing the ecological perspective on land and land management and looking for ways to reduce it to practice. The fact that all these projects seem to have been undertaken independently, with little or no knowledge of similar projects going on at the same time, reinforces the impression that these ideas were "in the air" at the time, at least among biologists and curators of botanical gardens. Seen from this perspective, these early attempts at ecocentric restoration were important, if generally underreported and underrecognized, events in the history of ecology because they represented attempts to reduce to practice—and so to realize in an especially demanding way—ideas such as the idea of the ecological community or the idea of succession to climax that underlay and to a great extent informed all these early projects.

The Desert Botanical Laboratory

Here a project undertaken in the course of creation of the Desert Botanical Laboratory (DBL) in Tucson, Arizona in 1906 is of special interest as a kind of ur-ecocentric restoration project, inconspicuous *as* a restoration project and in fact not identified—that is, realized—as one for an entire century. This project began in 1903 when the fledgling Carnegie Institution delegated two prominent botanists, Frederick Colville and Daniel MacDougal, to establish a laboratory in the arid Southwest. They sited it on 352 hectares comprising an intriguing variety of habitat types just west of Tucson.[2]

At first, their work focused on questions of environmental physiology. Soon, however, they were joined by noted University of Michigan forestry professor and conservationist Volney Spalding, whose severe arthritis had forced him to move to the desert.[3] At first, Spalding abandoned conservation and pitched in to help with the physiology research. But soon he realized the opportunities the laboratory offered for multispecies studies. In the spring of 1905, he began to study the plant species as community members in twelve habitat types, and in a letter to DBL director MacDougal he outlined what amounted to an emerging vision for a program of long-term ecological research.[4]

So far so good. However, the vegetation of the site had been severely degraded by five decades of grazing and browsing. So Spalding's thinking naturally turned to the question of its recovery. He proposed the construction of a triple-stranded barbed-wire fence around the entire site to exclude cattle, horses, burros, and goats. This was done in 1906, and noticeable recovery of the vegetation began immediately.[5] In 1930, Spalding's successor, Forrest Shreve, reported its success: "The long period without disturbance has brought the plant life back to virgin desert conditions such as one can find only in the most remote and ungrazed parts of Arizona."[6]

Although fencing may seem a minimalist form of restoration, it was, no less than the remeandering of a channelized waterway or the reintroduction of an extirpated species, a decisive act of compensation for or reversal of novel influences on an ecological system, which is a good, operational way of defining ecocentric restoration. Moreover, as far as motives were concerned, what Spalding and Shreve did at the DBL was exactly the same thing others were doing in an attempt to find out how to help degraded rangeland recover as rangeland, but they did it for an entirely different reason: not to improve the land as natural capital but simply to study the ecology of a desert association as it recovered from overgrazing. The contrast in aims here is neither subtle nor ambiguous. The work at the DBL entailed no mixture of motives comparable to those of scientists such as Arthur Sampson or Lincoln Ellison, who a few decades later attempted to restore historic grasslands in the Southwest because they thought they might be models for productive rangeland.[7]

This, then, was arguably an ecocentric restoration project, even though it entailed only the exclusion of a guild of exotic grazers and did not include deliberate reintroduction of extirpated species until many years later.

Importantly for us, however, because we are interested not only in the invention of ecocentric restoration but in its identification and realization as a distinctive form of land management, no one had thought of it that way when, in 2007, University of Arizona ecologist Michael Rosenzweig took over as director of the laboratory, now known as Tumamoc: People & Habitats. Rosenzweig initially envisioned developing the reserve as a site for research on the form of meliorative land management he has called reconciliation ecology.[8] This would have entailed tinkering with the community and the ecosystem in order to discover how a working arid landscape might also provide a rich array of habitats for species, including noneconomic species. But Rosenzweig has given a good deal of thought to restoration and its relationship to ecology and to other forms of land management over the past thirty years, and when he started looking into the history of the laboratory he began to realize that it had really been a restoration project from the beginning, that it was, as far as he knew, the only project with such a long history anywhere, and that unless he and his colleagues were prepared to compromise a unique situation resulting from a century of effort, they, as he says, "couldn't mess with it—at least not in that way."[9]

This was a discovery in its own right. It clearly illustrates the distinction between doing or finding something and discovering or inventing it. This 101-year delay between invention and realization was by no means unique to the DBL. In fact, it was typical—a pattern that would be repeated at project after project in the decades that followed, as we will see.

Vassar College

Whereas the DBL waited a full century to be recognized as an ecocentric restoration project, projects more conspicuous as restoration efforts got under way at a scattering of sites in the four decades that followed. The outcomes—or fates—of these initiatives differed widely, however, so that, taken together, they provide an excellent perspective on the development and realization of an idea whose time had not yet come.

Of these, the earliest we have identified was a project initiated in 1920 by Edith Roberts, a botany professor at Vassar College, as an outdoor laboratory and classroom for her students and, at the same time, a demonstration area for visitors. Roberts had studied with pioneering ecologist Henry Chandler Cowles at the University of Chicago, and the project clearly reflected Cowles's influence. Its aim, which Roberts described as "constructive conservation," was "to establish, on less than four acres of rough land, the plants native to Dutchess County, N.Y., in their correct associations, with the appropriate environmental factors of each association in this region."[10] The plan to limit the "collection" to associations native to the immediate area aligned this project with those at institutions in the United States such as the Brooklyn Botanic Garden and Garden in the Woods and distinguished it from the more cosmopolitan aims of those at sites such as Tervuren and Kirstenbosch. In addition, design considerations seem to have been limited to the questions of how to keep the project inconspicuous in the context of a landscaped campus and how to lay out access paths for use by students and visitors. In other words, ecology first, then the practicalities of research and education, and *then* aesthetics, a formulation that decisively reversed the priorities faced by the biologists working along parallel lines in the national parks a decade later. This inversion of priorities is nicely reflected in the fact that, as Roberts reports, the project itself, which was initially named the Dutchess County Botanical Garden, came to be known to the students involved as The Dutchess County Ecological Laboratory.

Besides this, the ecological condition of the site lent clarity and a certain conceptual edge to the project *as* a restoration project. Although it was possible for managers to regard lands in the national parks as natural as long as no hotels or parking lots were built on them and to think of manipulation in vague terms simply as "management", that was impossible on the Vassar site. No one would mistake its four acres of "rough land," apparently retaining little or nothing in the way of native vegetation, as natural, and any attempt to reestablish historic vegetation there would be unmistakably a restoration project. In fact, the project entailed a good deal of heavy lifting, including not only introduction of nearly all the appropriate species but also manipulation of topography and installation of an irrigation system to create exposure and drainage conditions suitable for the various plant associations.

At the same time, it is important to note that although the Vassar project was more clear-cut and conspicuous as a restoration project than the program conceived by the scientists at the DBL fourteen years earlier or by the Park Service biologists a dozen years later, it was actually less holistic, so to speak, because it focused exclusively on plants. Besides this, Roberts used existing natural areas rather than records of historic vegetation in defining objectives for her plantings. And the plantings did not take into account dynamic features such as succession and the development of habitat for animals that Emily Griswold sees as distinguishing ecogeographic displays from ecological restoration.[11]

Overall, the project was quite successful. In 1933, Roberts reported that "twenty-eight of the thirty associations of the county are so well established that fifty per cent of their plant members are at hand for taxonomic, morphological and physiological studies." Records of the project indicate that at that time 93 percent of the roughly 2,000 plant species on Roberts's list were established in the plantings, and a 1948 newspaper account noted that at that time only three were missing.[12]

What set the project at Vassar apart from earlier ecogeographic displays was its decisive commitment to creation of examples of the historic vegetation of an area explicitly for ecological study, and with a studied, if diplomatically expressed, indifference to their value as displays or elements in an amenity landscape.

This was still gardening, but it was ecological gardening and made few if any concessions to the tastes, expectations, or convenience of visitors. The resulting bits of vegetation were no more displays than the plantings of alfalfa or hybrid corn at an agricultural experiment station are displays or a flat full of plants in a genetics experiment is a bouquet. This made it different from Olmsted and Vaux's plan for the Niagara Reserve because it explicitly rejected the utilitarian aims even of aesthetics and reflected a concern for—or interest in—the various plant associations for their own sake, as objects of scientific curiosity. At the same time, in contrast with the Niagara Reserve project, it was carried out on a very small scale and with no idea of creating dynamic, self-sustaining ecosystems that might serve not only as examples of their natural counterparts but as models, the creation and management of which might be directly relevant to conservation on a landscape scale.

This was also true of a similar project undertaken, apparently almost as an afterthought, in connection with the creation of an arboretum at Carleton College in Northfield, Minnesota on some 140 hectares of land adjacent to campus. In 1926 botany professor Harvey E. Stork, noting that only four arboreta "of any importance" existed in the United States at the time and pointing out the variety of habitat conditions offered by the site, which included upland and bluffs as well as a small creek and a 3-mile stretch of the Cannon River, proposed creation of an arboretum.[13] Stork conceived the project in traditional terms, as "a museum of trees and shrubs" and "a proving ground for new materials of landscape gardening."[14] But early on he and groundskeeper D. Blake Stewart began collaborating on a planting of native trees and shrubs in one area. As they added understory and ground-layer species rescued from nearby properties undergoing development, the area came to resemble the mesic hardwood forest of the region, at least with respect to the species and to some extent the distribution of the vascular plants.[15]

The Holden Arboretum

What we are seeing here is not a historic watershed, the decisive crossing of a conceptual boundary, so much as a kind of rapids in which an idea tumbles through a series of small barriers and declivities as it takes shape in a certain direction, guided not so much by a brilliant insight, a flash of genius, or a grand vision as by a certain instinct or gravity, at times following and at times resisting the lay of the land.

With this metaphor in mind, we may move downstream a few years, to what was, as far as we know, the first project to link the objectifying, ecological commitment reflected in the Vassar project with the commitment to wholesale assembly of an actual ecosystem. This project was launched in 1930 on the first parcel of land acquired for development of the Holden Arboretum near Cleveland. Here again, as at Vassar a decade earlier and as would occur at the UW–Madison Arboretum a few years later, the project was undertaken under the auspices of an educational institution, in this case the Cleveland Museum of Natural History, and was conceived and carried out by people with biological rather than horticultural or landscape design interests. The key figure in the early years was Benjamin Patterson Bole Jr., a small-mammal expert with broad ecological interests. Like both of the other projects, it began with a commitment to make ecological and historical accuracy trump utilitarian interests. But this would be carried out on roughly 40 hectares, a full order of magnitude larger than Edith Roberts's "four acres of rough land," and was ecologically more comprehensive in paying at least some attention to animals, specifically birds and small mammals.

Taken together, these commitments meant that the resulting ecosystems, as envisioned, would have at least some of the character of their natural counterparts as they developed and interacted on a landscape scale and so might have implications for land management and conservation beyond their value as educational displays. The much better-known project undertaken at the UW–Madison Arboretum just a few years later is often identified as the first such project. But the project at the Holden Arboretum, though it was largely abandoned within a decade and passed into obscurity, not only reflected essentially the same idea but made significant progress toward its implementation.

The planners themselves clearly recognized the novelty of the project, although they expressed it rather modestly. Making it out to be less a bold initiative than an opportunity and even an obligation arising from the emergence of the new science of ecology, they noted that "since the modern basis of plant study tends more and more to rest upon ecology, such a study is very important," and that "the Holden Arboretum might well choose, as its principal contribution to the knowledge of plants, a complete presentation of plant successions in their broad biologic aspects."[16]

The aims of the project, as elaborated further by museum ecologist Arthur B. Williams in a progress report three years later, were unmistakably what we are calling ecocentric restoration:

> Eventually we hope to establish an outdoor ecological museum by reproducing as many of the plant communities of the temperate parts of the globe as we can. Each forest type will be made as complete in every detail as climate, soil and experience permit; and eventually it will be possible, we hope, to greatly improve the latter two factors, and perhaps even the first with the aid of a greenhouse. Special emphasis will be laid on North American types. We will eventually be able to show a plant, a shrub or a tree in its proper relationships with other species of its native land, and not merely in a group of taxonomically related species from all parts of the world.[17]

Though much larger than Edith Roberts's small plot, the 40 hectares available for the project was tiny compared with a national park. As a result, the project was small enough to manage intensively yet large enough to suggest the possibility that some of the created associations might develop a measure of ecological autonomy. One, a re-created bog, actually did, a number of its distinctive species surviving nearly a half-century of neglect after the virtual abandonment of the project in the 1940s. In a 1935 letter summarizing some of the principles underlying the project, Harold L. Madison, director of the Cleveland Museum of Natural History, made it clear that those involved were thinking of the ecosystems they were putting together as relevant to conservation on a landscape scale, but he also noted that they were too small and too close together to maintain themselves.[18]

As far as the ecosystems to be created were concerned, the Holden Arboretum biologists aimed at restoration of a wide range of types, including ones found in the mountain West, the Pacific Northwest, and even parts of Europe and Asia where conditions are similar to those in the Midwest. As Williams noted, however, the emphasis was on North American types, and indeed on those of northeastern Ohio. Associations on which Bole and his colleagues actually made progress in the decade or so in which the project proceeded in a consistent way included restoration — or creation — of examples of the several successional stages then thought to be characteristic of the beech–maple–hemlock complex of community types as well as oak–hickory forest, river bottom forest, swamps, and several aquatic systems.

This included removal of species such as maples, to move a maple forest back to an earlier successional stage; introduction of plant species, including ground-layer and understory as well as canopy species; and creation of several artificial basins for aquatic associations. Interestingly, reflecting a clear sense that creation and maintenance of these communities would entail resisting as well as taking advantage of community dynamics, Harold Madison noted in his introduction to the 1935 progress report for the project that "since almost every plant association is steadily and relentlessly being crowded out by a more dominant association the practice would be to protect and preserve each association from invasion by its stronger neighbors."[19] Complementing the emphasis on plants, and presumably reflecting Bole's interest in animals, the early reports included censuses of the birds and small mammals occupying the restoration and creation areas in successive years. Early on, demonstrating their commitment to research and to development of the ecosystems being re-created as "ecological laboratories," the project managers set up a series of 100-foot-square quadrats, marked, a bit ironically, with posts made from specimens of American chestnut killed a few years earlier by chestnut blight.

Equally important for us, the biologists prevailed over the designers, at least at first, by a combined strategy of deferral and segregation. The early plans stated explicitly that, although the aim was "eventually" to create "classified" (i.e., taxonomic) collections "as other arboretums have done . . . the prime emphasis . . . will be on the natural associations of the plants rather than on . . . taxonomic relationships or horticultural possibilities, at least so far as the present property is concerned."[20] Consistent with this emphasis, by far the larger part of the property was to be devoted to the ecological plantings, with only a small area being designated for plantings of ornamental shrubs.

However, the perennial tension between design and reproduction existed at the Holden as much as in the national parks. When Elmer Merrill, the director of Harvard University's Arnold Arboretum, visited the site in 1938, he wound up chiding the staff for devoting too little attention to the development of ornamental and landscape plantings, noting that "it is the outstanding horticultural features of the Arnold Arboretum that have developed local interest in the institution—or I should say national and international interest, and what is more important, the necessary financial support." Merrill pointed out that it was not the Arnold's large library or herbarium, and not even its extensive collection of trees, but rather the displays of "Oh, my! flowers" that most impressed visitors. He added that "when one considers 40,000 *pedestrians* visiting the lilac displays in one day (in this age of stepping on the gas), one must realize that such a display has an enormous drawing power"—more, presumably, than the Holden's hemlock forests, reconstituted bogs, and swamps were likely to have. He went on to advise that as soon as possible "attention should be given to mass and display plantings of such groups as lilacs . . . oriental crabs and cherries, azaleas and rhododendrons, and oriental and occidental dogwoods." He concluded this advice with an offer of seeds and cuttings from the Arnold's collection.[21]

Such advice did not fall on deaf ears. Brian Parsons, who has worked at the Holden since 1977 and who played a key role in the rediscovery of the early plantings and the revival of the restoration effort, notes that advice such as Merrill's resulted in some disagreement among the directors and ultimately led them to move away from Bole's ideas. By the end of the decade, paralleling developments at Yosemite during the same period, the Holden's ambitious restoration efforts had begun to be superseded by projects held to have more popular appeal. A 1939 report notes that among developments of that year that would "combine to make that year historic in Arboretum history" were plantings that "for the first time" extended "the Arboretum's program into the systematic and ornamental field."[22]

Down Under

The scientists in Ohio were apparently unaware of it, but they were not the only ones attempting to re-create whole, stand-alone ecosystems. Harvey Stork and Blake Stewart had been tinkering with their small-scale restorations in Minnesota for several years by the time the project at the Holden got under way. And when we went looking for early projects, mainly by asking colleagues from different parts of the world for leads, we came up with a number of "firsts"—Kitty Hawks of a sort—all undertaken around the same time, in the early and mid-1930s. Although as far as we can tell these were undertaken independently, the bright idea or hobby horse of one or two or a handful of individuals rather than an idea spread around in journals or even by word of mouth, they had strikingly similar aims. And in fact, just about the time workers were setting up their plantings in Minnesota, Ohio—and, as we will see, Wisconsin—two strikingly similar projects were getting under way on the opposite side of the world, in southern Australia. Both paralleled the projects in the United States in important ways, and like them they pioneered uses of restoration and ways of going about it that would not come into general use for many decades.

In 1935, more than a decade before Aldo Leopold wrote of "two middle-aged farmers" rising before dawn to transplant tamaracks to their farm in defiance of a popular prejudice against tamarack,[23] a New South Wales dairy farmer named Ambrose Crawford, then fifty-four, became concerned about a plan to clear and "grass" a 1.7-hectare remnant of subtropical rainforest dominated by the white booyong (*Hieritiera trifoliolata*) at Lumley Park in Alstonville. Looking for a way to rescue this remnant, he convinced the shire council to designate it a "Preserve for Native Trees."

Restorationist Tein McDonald, who is editor of the Australian journal *Ecological Management & Restoration*, drew our attention to the Lumley Park project and also to a project under way at the same time in nearby Broken Hill, New South Wales. She then took the trouble to look into their history, exploring records held by the Alstonville Plateau Historical Society and also interviewing Crawford's daughter, Dorothy, who has clear memories of her father's weekend efforts on behalf of the forest remnant at Lumley Park. What she found was that both stories offered, in different ways, striking parallels both with the landscape history and with the restoration efforts we have been exploring in the United States, all of which were undertaken at about the same time and which are the earliest examples we know of ecocentric restoration in its fully developed—if not fully realized—form.

Ambrose Crawford, like his counterparts in the United States, had grown up in a landscape undergoing large-scale transformation. The 75,000-hectare white booyong forest on the Alstonville Plateau in northern New South Wales, known as the Big Scrub, had been cleared almost entirely between the 1860s and the end of the nineteenth century, and only scattered remnants remained. Crawford, born in 1880, experienced this transformation firsthand at just the time it reached its conclusion early in the twentieth century. This gave him a personal connection with the all-but-vanished ecosystem, and it seems that the tiny remnant in Lumley Park was his version of the scattered bits and pieces of bog or tallgrass prairie that were inspiring some of the earliest restoration efforts in the United States at just this time. Crawford described his interest in the rescue and conservation of such relics of the past in a brief account of the project that he wrote years later, noting that the bit of forest at Lumley Park "had never been 'felled' except for cedar, and was a fair sample of what the 'Big Scrub' was like before the Settlers came here to make their permanent homes from 1865 and onwards."[24]

What Crawford and his colleagues were dealing with at Lumley Park was in fact a reasonably intact remnant—a sacred grove of sorts—that called first for protection and led Crawford's team into restoration only gradually as they realized this was necessary to ensure the survival of the remnant forest. This actually—and significantly—made the Lumley Park project a miniature version of the restoration efforts being proposed for the national parks in the United States—active, ongoing restoration of an area that most regarded as an original, natural, or pristine ecosystem not needing restoration.

Crawford soon realized this was not the case, however, and undertook a program of restoration that occupied him for most of the rest of his long life. This entailed intensive clearing of weedy invaders and reintroduction of plant species appropriate to the site or to Big Scrub ecosystems generally. To get this work done, Crawford corresponded with botanists at the botanical gardens in Sydney and Brisbane and eventually became expert on the local flora. He also, in keeping with widespread practice among conservationists at the time, recruited friends, who began spending Saturdays working at the site. This set the project apart from projects such as those in the United States, which were carried out almost entirely by professionals. And it anticipated the emergence of volunteer-based restoration efforts, which became a distinctive feature of the "restoration movement" in the United States decades later. From our perspective the results seem representative of what McDonald nicely calls the "less articulate" beginnings of what has since evolved into a distinct land management paradigm. She notes that "the site still looks like a remnant" and, despite its small size, "is home to three threatened plant and two threatened animal species."

If the tiny Lumley Park project was an early example of volunteer-driven restoration, another project undertaken in Australia around the same time took the important step of restoring a high-quality swatch of native vegetation and then expanding it to landscape scale. McDonald suggests that this project, spearheaded by field naturalist Albert Morris, was "probably the first ecological reconstruction project in Australia using the local ecosystem as a reference." Working in the same ecologically transformed landscape as Ambrose Crawford, Morris had expertise in the native flora and, like John Weaver in the United States, had made a close study of the response of the native vegetation to disturbances such as the severe droughts that occurred in southern Australia, as in the Midwest of North America, during this period. Over the years Morris had developed the idea that the primary cause of vegetation decline in the region was not drought but overgrazing, which could be reversed by fencing to exclude sheep and rabbits. This was basically the idea that had led to the fencing at

Arizona's DBL, with the significant difference that Morris planned to reintroduce native species he believed would not be able to recover on the site naturally. It also reflected the idea, consistent with the succession-to-climax model of association dynamics that prevailed at the time, that, with some help, the ecosystem had the ability to heal itself. Morris argued that "this country is not a desert . . . it is a huge garden." And he insisted that if the land could be brought back to its original condition, the garden would soon replace the desert.[25]

Professionals and professors scoffed at this idea. But in 1936, with southern Australia in the midst of its own Dust Bowl, Zinc Corporation, a mining company in Broken Hill, New South Wales, desperate to find a way to reduce blowing dust that was rendering parts of the town virtually uninhabitable, commissioned Morris to try out his idea on a 9-hectare tract adjacent to the town. This proved so successful that two years later the project was scaled up to include a zone a kilometer wide extending around half of the town.

This, too, was successful, and botanists documented dramatic recovery of vegetation on Morris's "regeneration areas" after the drought broke in 1939. Unfortunately, Morris died that year. But his project was important not only as a demonstration of the restoration and recovery of an ecosystem under intelligent restorative management but as an early instance of restoration as a conservation strategy. Those who undertook the other restoration projects we have profiled seem to have regarded them as experiments or demonstrations with little direct relevance to the conservation of actual ecosystems or landscapes. In contrast, Morris clearly had conservation in mind all along. He set up his first project as a pilot project, and he scaled it up by several orders of magnitude as soon as the results confirmed his ideas.

This was an important—indeed crucial—step in the realization of the value of restoration. At the same time, it forces us to consider the motives driving the project. Those of the Zinc Corporation were, presumably, primarily utilitarian. But Morris's were not, and McDonald, having examined the record in some detail, reports that it is her impression that he may well have been at least as interested in the reestablishment of the native vegetation as in dust abatement. She notes his long-term interest in the native plants of the region and the fact that, although he had used some exotic species in earlier plantings, he used none in the two projects he undertook for Broken Hill. (Arthur Sampson and Lincoln Ellison, as we have seen, moved in just the opposite direction in their revegetation efforts in Utah.) She acknowledges that he argued that the native vegetation would be best suited to solve the dust problem, but she suggests that he probably would have had to use this argument to make the case for sticking to native species. In such a "real-world" setting, a mix of motives was inevitable. And the trick of connecting an ecocentric restoration project with some "practical" objective in order to justify it in utilitarian terms has since become standard practice.

Wisconsin

Of the half-dozen or so projects we have identified that qualify in most respects as early or pioneering attempts at ecocentric restoration, by far the best known is the project undertaken at the University of Wisconsin Arboretum in Madison in the mid-1930s. This project has gained prominence for a number of reasons. One is its scale, roughly 200 hectares, several times larger than the runner-up Holden Arboretum project. Another is continuity, the restoration effort having been carried out continuously, though at varying levels of intensity, down to the present. A third is the clarity with which the aims of the project were articulated at the outset, and, not least, the fact that Aldo Leopold, perhaps the most prominent conservationist of the first half of the twentieth century, played a key role in the project in its first few years and articulated the philosophy behind it in published writing that has been among the most influential environmental writing of the twentieth century. Yet another reason was that in the late 1970s several members of the arboretum's staff began to draw attention to the early restoration efforts carried out there and to explore their implications for conservation.

For all these reasons, this project is consistently identified as the point of origin of environmental restoration. This, it should be clear by now, is an oversimplification, the sort of mythologizing that happens naturally when a discipline takes shape and needs an origin myth. Nevertheless, the story of the formative years of the UW–Madison Arboretum provides a fascinating case study, illustrating not only an important milestone in the invention of ecocentric restoration but also the difficulty its proponents and detractors alike have had in recognizing its distinctive character and realizing its distinctive benefits. It is especially interesting because Leopold, who played a key role in launching the project, was developing his ideas about the intrinsic value of nature at just the time the arboretum project was taking shape, so this experiment in ecocentric restoration provides a valuable opportunity to consider what Leopold himself made of this idea and how far he was willing to push it in practice.

The story of the UW–Madison Arboretum began in the late 1920s, when a handful of civic leaders who had been attempting to raise support and funding for creation of a public park on land bordering small Lake Wingra on the western edge of the city changed their objectives and began promoting the project as an arboretum for the university.[26] This shift in aims, from public park to "outdoor laboratory and classroom," is crucial for us because it meant that the arboretum, "wildlife refuge," or "forest experiment preserve," as it was variously labeled, would serve primarily as a field station for a handful of biology professors at a major land grant university rather than as a public park or pleasure ground. In other words, it would be largely free of the political and institutional pressures that stymied similar initiatives in at least a few national parks at exactly the same time. It would be carried out under conditions much like those Edith Roberts enjoyed at Vassar College, though on a scale two orders of magnitude larger. At the same time, like the Vassar project, it explicitly privileged research and education over recreational and amenity concerns. Leopold made this clear at the outset, noting in the short talk he gave at the dedication of the arboretum in June 1934 that the project "will be done for research rather than for amusement" and "for use by the University, rather than for use by the town."[27] Here he officially cut the project off from its roots in a civic drive to create public green space and warned off anyone who imagined the project evolving into a pleasure ground for the laity.

In many respects, the project was carried out in a kind of middle ground between the conditions Roberts enjoyed at Vassar and the conditions that ultimately frustrated the efforts of George Wright and his colleagues in the national parks. Besides being an academic project under the control of biologists with a mandate for research and education and no direct pressure to serve pleasure-seeking visitors, the project was, like the one at Vassar, to be carried out on dramatically altered land. Much of the upland had been cultivated for about three quarters of a century and consisted of a patchwork of derelict pastures, overgrazed woodlots, and recently abandoned cropland. Extensive lowlands around the lake had been subject to dredging and channeling in an abortive attempt at residential development.[28] The property, which by the mid-1940s included some 440 hectares, was small compared with a park such as Yosemite or Shenandoah. But it was large enough to allow for plantings — or restorations — on a scale that, given the ecological thinking at the time, the project planners might reasonably suppose would allow the restored ecosystems a measure of ecological autonomy and what Leopold would later call integrity. It was large enough that the planners applied for and obtained a unit of the Civilian Conservation Corps (CCC) to carry out the early work on the property, including the large-scale gardening effort that would be needed to restore a hundred or so hectares of degraded land. The first CCC "enrollees" arrived in August 1935, and soon "Camp Madison," in the heart of the arboretum, was temporary home to more than a hundred young men and was locally distinguished as "the only CCC camp on a university campus."[29]

These conditions would be clearly reflected in the plan for the property, developed by a group of faculty and administrators during the years immediately following acquisition of the first land in 1932. The aim, which Leopold summarized in a speech at the dedication, would be "to reconstruct, primarily for the use of the University, a sample of original Wisconsin—a sample of what Dane County looked like when our ancestors arrived here during the 1840s."[30]

This was almost identical to the plan the group at the Holden Arboretum had developed just a few years earlier. Like their contemporaries in Ohio, the UW group focused on the grassland and forest associations native to the region but allowed some space for plant community types that flourish several biomes away, notably the pine and spruce forests of northern Wisconsin and forest types of the Ohio River Valley. More than their counterparts working on other projects, they were committed to historic accuracy and drew, as the project took shape, on surveyors' maps and other sources of information about the historic ecosystems they aimed to restore. Their objectives, like those at the Holden, were ecologically comprehensive and eventually included processes—notably the burning of vegetation—and the introduction, and when necessary control, of animal as well as plant species. This reflected the influence of Leopold, who, by the time the arboretum planners sat down to ponder ends and means for the new project, had gained, on top of his years of work as a forester, a reputation as a leader in the new discipline of game—or wildlife—management. Leopold's seminal book *Game Management* was published in May 1933, and just two months later he accepted an offer from the university to assume what would be the first professorship of game management in the country. Not surprisingly, this new and rapidly developing interest was reflected in Leopold's early contributions to the planning for the project. In fact, game animals were prominently represented in early plans for the arboretum, and the emphasis soon expanded to include attention to nongame species as well. At the same time, plants were well represented by botany professor Norman Fassett, a plant taxonomist who had an intense interest in the flora of Wisconsin and its conservation and whom Leopold biographer Curt Meine describes as "Leopold's equivalent in the university's botany department," noting that he undoubtedly influenced Leopold in many ways. Fassett clearly stimulated Leopold's interest in the regional flora.[31] Indeed, Frank Court, who has done extensive archival research for a history of the arboretum, stresses that the record makes it clear that no one person was responsible for the ideas that took shape in its early years. He notes that the idea of restoring a prairie, which became both the ecological and the visual centerpiece of the arboretum's collection of restored communities, had been articulated by at least four of those involved by 1933 and that "the record shows that the idea was in the air and had been for years" before the dedication in 1934.[32]

In 1935 two of Fassett's students, John Thomson and Roger Reeve, carried out the first planting trials on the abandoned pasture chosen for restoration of prairie.[33] Others, from disciplines ranging from botany and zoology to soils and limnology, also participated in the early development of the arboretum, providing a range of expertise reflected in the commitment to include "all the parts" that characterized the plan for the arboretum's restoration effort from the start and has guided the work there ever since.

Equally important in the planning mix was the role of the designers — in this case, principally horticulture professor G. William Longenecker, who joined enthusiastically in a project he envisioned as the creation of a "native America in miniature."[34] Longenecker played an important role in the development of the arboretum but one that contrasted sharply with the role of landscape architects in the national parks, where the architects consistently had the upper hand. At the arboretum, in contrast, the two parties seem to have worked well together, in part, it seems, because they maintained a wall of separation between the aesthetic interpretation of landscape and the attempt to re-create, reproduce, copy, or restore accurate representations of objectively defined ecosystems. Thus Longenecker did design work on various amenity features of the arboretum, such as entries, visitation areas, and several formally laid out collections of trees and shrubs. These provided the arboretum's version of Elmer Merrill's "Oh, my! flowers" for visitors who wanted to see lilacs or crabapples in bloom. (Oral tradition has it that he also helped plan the layout of the collection of restored communities, setting the pine and spruce forests around the "central prairie," in a lazy-Susan pattern that made it possible for students and visitors to take a kind of visual tour of midwestern vegetation from a single spot on the prairie. Interestingly, even this modest intrusion of design considerations resulted in some ecologically inappropriate juxtapositions, most conspicuously the placing of northern pine and spruce forests adjacent to a tallgrass prairie, rendering the arboretum's collection somewhat quaint from the perspective of landscape ecology.) But at the community level, which was the focus of the early thinking at the arboretum, Longenecker played no role at all, at least as a designer. The biologists went about their business in the backcountry of the property, focusing on ecological authenticity.

When the time came to begin systematic work on the prairie, the planners cast about for a prairie ecologist to plan and supervise the work. They wound up hiring Theodore Sperry, a young ecologist who had recently completed a Ph.D. at the University of Illinois, focusing on prairies. Sperry arrived at "Camp Madison" early in 1936 and supervised the prairie project through its first five years, establishing a precedent for development of the ecological collection by ecologists that has prevailed ever since.

Whereas Ted Sperry mostly ignored aesthetic considerations in stitching together his prairie, making his plantings in an irregular patchwork of single-species plots, botanist Henry Greene exercised his highly developed sense of prairie aesthetics in his efforts to restore the arboretum's second large prairie in the late 1940s. However, he did this not in the interest of developing a prairie "motive" in the manner of architects such as Jensen and Miller but as a sensitive guide to accuracy in the selection of species and distribution of plants. Indeed, Greene had little interest in making an impression on the public. He undertook the project with the understanding that the site would be *protected* from visitors, and it was two decades before he acceded to mowing of a path through what had by then developed into one of the most successfully restored prairies anywhere. Working in self-imposed isolation on a back 40 out of sight of any road and under an agreement that essentially precluded public access, Greene was about as far as you could get from the conditions that prevailed in the national parks, where the aims of restoration were consistently subordinated to those of aesthetic enhancement of the landscape and encouragement of watchable populations of large animals. In general, design and restoration were segregated efforts at the arboretum. Although Sperry made some concessions to public taste in setting out the first plantings on the prairies, all concerned understood that these were short-term expedients, tactics calculated to reassure visitors who might entertain some skepticism about what was clearly an unconventional approach to the development of an arboretum.[35]

Landscape historian Philip Pauly recently argued that the restoration efforts at the arboretum—the early work on what would later be named

A Prairie Proving Ground

When Aldo Leopold responded to Norman Fassett's suggestions by penciling in a site for prairie amid a scattering of food patches for game birds on a rough map of the UW–Madison Arboretum in the fall of 1933,[a] he was, probably without realizing it, sounding a keynote for the practice of ecocentric restoration as it would take shape over the following decades. What Leopold labeled "Central Prairie" would quickly become the centerpiece of the arboretum's collection of restored communities, which eventually included some two dozen community types, and this pattern would be repeated over the years as the tallgrass prairies of the North American Midwest became the cradle, the proving ground, and eventually the poster child of ecocentric restoration.

There are a number of reasons for this. Perhaps the most obvious is the rate, scale, and thoroughness with which the prairies had been plowed down, an entire subcontinent converted to cropland and settlements in the span of a single lifetime. Europeans had changed New World landscapes wherever they encountered them, of course, but nowhere else so rapidly, so completely, and on so large a scale.

Besides this, in marked contrast with the eastern forests, which grew back on cleared land once it was abandoned, prairies, once plowed down, did not recover even a semblance of their original composition or character. Even those not destroyed outright by the plow soon vanished, as trees occupied them in the absence of fire. Eventually, those who were paying attention realized that if prairies were to survive into the future, it would be necessary to replant them.

Fortunately, far more than forests, the prairie lent itself to such treatment. Fine textured and reproducible on a small scale, its vegetation composed mainly of long-lived herbaceous species, most of which are easy to handle and reach maturity quickly, the prairie proved amenable to horticultural and even agronomic techniques.

Perhaps most important, as far as the distinction between ecocentric and meliorative restoration is concerned, although Americans tended to romanticize the prairies, they had little taste for or interest in actual prairies and no economic incentive to conserve, much less restore them. Settlers had come to the Midwest to plow down grass, not admire it, and for their descendants tallgrasses other than corn and wheat signaled neglected land. In contrast with replanting a forest, restoration of a prairie would be seen not as an economic gain but as an economic sacrifice — the economic equivalent of closing a factory while creating a landscape most would see as a weedpatch.

A Prairie Proving Ground
Continued

Partly for this reason, prairies have never been the subject of laws or regulations requiring either their protection or their restoration, as wetlands and surface-mined lands eventually would be.

This being the case, prairie restoration would necessarily be a labor of love, motivated by concern for or interest in the biome itself rather than in a desire to improve the land for some economic purpose or the need to comply with laws and regulations.

Aldo Leopold might appeal to the value of prairie in building soil, as he did at the dedication of the UW–Madison Arboretum. But the claim that to restore soils degraded by farming it would be necessary to restore prairies whole, including all their species, obviously pushes the idea that the functioning of an ecosystem depends on its species composition to, if not beyond, its limits. Research over the past couple of decades has shown that soils degraded by agriculture do indeed recover carbon, nitrogen, and crumb structure under restored prairie. However, it is also clear that this process is affected by many interacting factors, including soil conditions, climate, and vegetation. Mike Miller and Julie Jastrow, who investigate this process at Argonne National Laboratory near Chicago, point out that the most important factor is probably the amount and chemistry of the carbon residues the vegetation puts into the soil and that the rarer species probably play correspondingly minor roles in the process.[b]

---

a. Aldo Leopold, "University Arboretum Wild Life Management Plan," University of Wisconsin–Madison Archives. Aldo Leopold papers, 38/00/6, Box 2.

b. Interviews with Jastrow and Miller, March 14, 2011. See Sarah L. O'Brien, Julie D. Jastrow, David A. Grimley, and Miquel A. Gonzalez-Meier, "Moisture and Vegetation Controls on Decadal-Scale Accrual of Soil Organic Carbon and Total Nitrogen in Restored Grasslands," *Global Change Biology* 16 (2010): 2573–88; R. Matamala, J. D. Jastrow, R. M. Miller, and C. T. Garten, "Temporal Changes in C and N Stocks of Restored Prairie: Implications for C Sequestration Strategies," *Ecological Applications* 18, no. 6 (2008): 1470–88.

Curtis Prairie in particular—was never more than a pretense. He notes, for example, that of the several hundred species of flowering plants known to have existed on the old prairies, Sperry introduced only about 50 during the five years he supervised the early plantings and that in some cases he favored attractive, showy species and left out less conspicuous or unpopular species such as poison ivy and the appropriate native species of nettle and thistle. This, however, is a serious misunderstanding of what actually went on at the arboretum and the thinking and intentions behind the work there. In a 1935 report Werner Nagel, assistant regional wildlife inspector for the National Park Service, noted the effort being made at the arboretum to ensure that the "original prairie" is "accurately reproduced."[36] Sperry himself noted in a 1939 report on the prairie restoration project that the plan for the arboretum was "to include eventually all native vegetation types, and in so far as possible, all plant species characteristic of these types," and added that "a large number of animal species are likewise being given considerable attention and all possible encouragement."[37] Pauly's characterization of the project is based on records of just the project's first few years and overlooks work carried out after 1941. Planting records show that hundreds of species, including many of the "bad actors" Pauly identified as having been omitted, were introduced into the prairies in the decades that followed — botanical versions of Augustine's mice or the stones Sperry had his "boys" rescatter on the site in pursuit of postglacial authenticity. Those not brought in deliberately were ones, such as poison ivy, that were already there or could be counted on to come in on their own.

Beyond nettles and thistles, however, an even more dramatic contradiction of the notion that the arboretum team favored the more congenial elements in the ecosystems they undertook to restore was the introduction of fire. As we noted in chapter 1, the use of this powerful land management technology predated even the emergence of our species. Fire, both natural and anthropogenic, had played key roles in the shaping of many New World ecosystems, but it did not figure in the vision Americans entertained of a land they tended to regard as unspoiled nature, fresh from the hand of the creator.[38] Demonized as the very emblem of destructive power, fire certainly upstaged nettles and thistles as a "negative" in the popular imagination.[39] And yet when the early attempts at prairie restoration seemed to be failing, the arboretum developers promptly began burning the plantings, emulating the fires that had swept the prairies for millenia prior to European settlement.

The results were spectacular. Frequent burns proved to be the key to the restoration of prairies in many situations.[40] And this rediscovery of an ancient technology not only opened the way to the restoration of prairies but was also a step toward the rediscovery of the role humans had played in creating the ecosystems Americans had regarded as natural.

Such violations of conventional sensibility actually set the arboretum project (and its sister projects at other sites) apart from the mainstream conservation of the time in important ways. Leopold made this point emphatically in a February 1940 note to Paul Brown in the Omaha office of the National Park Service, which administered the work of the CCC at the arboretum. Addressing Brown as "one of those few who have understood the Arboretum idea," Leopold, as though inserting himself into the argument recently lost by George Wright and his colleagues at Yosemite, objected to the transfer of budget funds from wildlife research to construction and urged that the Park Service recognize the arboretum as "a special case." "I think," he wrote, that

> I detect in this proposal a basic conflict between two opposing viewpoints. The N.P.S. has developed, for use in ordinary parks, a set of "canned" procedures which give priority to construction. For ordinary parks there is doubtless reason for this. When forcibly applied to the Arboretum, however, these procedures simply do not fit. They will force us to abandon certain wildlife work which I regard as important, in favor of certain constructions which to my mind are of very minor importance. If the Arboretum becomes an ordinary park, my interest in it will cease.[41]

Leopold concluded with a comment that suggests he was not aware of the struggles of his colleagues working in "ordinary" parks such as Yosemite. "I venture to guess that the N.P.S. camps on the national parks are not dropping wildlife work to raise funds for construction. If so, then we have indeed become an age of engineers."

If this sounds a bit elitist, it is precisely this exclusionary principle, setting other species before people, that distinguishes ecocentric restoration from meliorative land management and that distinguished the arboretum, as Leopold and his colleagues conceived it and valued it, from "ordinary" parks.

By the time Camp Madison closed, on the eve of Pearl Harbor in November 1941, the basic outlines of the arboretum project had taken shape. Developed over the decades since along the lines sketched out by the planners, the arboretum is, as far as we know, the only project of its kind to have been pursued so consistently for so long, and it has gained a reputation as the *locus classicus* of ecological restoration, with Aldo Leopold being credited as the genius behind it.[42]

Environmental philosopher Eric Higgs has objected to this characterization, noting, like Marcus Hall, that environmental restoration has a long history and that it means different things to different people.[43] Fair enough, as long as we are talking about restoration in a very broad sense — indeed restoration of *something*, the nature of which is always up for debate: a transitive verb without its direct object, meaning nothing, sug-gesting all good things.

That is not what we are exploring here. We are not interested in the long history of people's attempts to create and maintain habitat for them-selves. We are interested in the invention and realization of a distinctive form of land management: restoration of ecosystems, defined in ecological terms and including all their parts and processes, whether we happen to re-gard them as good or not. That *is* a new idea. It is also an idea that takes shape pretty naturally once the key elements — concern for the old ecosys-tems, a sense of historical time, perhaps a bit of nostalgia, the idea of resto-ration as redemptive, and ecology — come together. As we have seen, this did not happen, as with Archimedes in his bathtub, in one flash of insight. Rather, it came to many people facing an altered landscape with con-cern — a botany professor in New York, a groundskeeper in Minnesota, a dairy farmer in Australia, and groups of scientists in Arizona, Ohio, and Wisconsin.

The Wisconsin project has emerged as something of an origin myth for a number of reasons, as we have mentioned. Perhaps the most salient of these is that it acted out the new idea so clearly, its prairie and forest resto-ration efforts contrasting so sharply, and so self-consciously with the melio-rative land management projects being carried out all around them.

Fair enough. Every enterprise needs a mythology to give it identity, meaning, and purpose. At the same time, it would be a mistake to attribute the invention of ecocentric restoration exclusively to that project, much less to one person. For one thing, it is clear, even from our (certainly in-complete) list of early projects, that it was not unique and was not even the first such project. For another, it involved the collaboration of a number of people, not just the bright ideas of one person. And, most important, those who launched the project fell far short of realizing its distinctive value in the dimensions of conservation, research, education and learning, expres-sion, and meaning making.

It is no discredit to those pioneers to point out that they did not realize what has been realized and reduced to practice by hundreds, even thou-sands of practitioners, investigators, and observers in the three-quarters of a century since.

In any case, because what we are interested in here is not only the in-vention of ecocentric restoration in a technical sense but also the discovery and realization of its distinctive value, this is an important matter for us, and we will explore it in some detail in the chapters that follow.

Chapter 5

Neglect

The closing of "Camp Madison," though an event of little interest beyond those directly involved, was nevertheless a kind of punctuation, marking the end of a chapter in the history of ecocentric restoration. During the previous decade the notion of ecocentric restoration had been, we may say, "invented." It had been put on the ground at a handful of sites by practitioners who, to varying degrees, recognized it as a novel variation on the ancient practice of environmental management and stewardship. Moreover, a few of these, such as those at Vassar College, Broken Hill, and the Holden and UW–Madison arboreta, were fairly conspicuously placed, at least in conservation circles. Like a new-model car or clothing fashion, the idea of ecocentric restoration was on the shelf, ready for rollout, marketing, and implementation.

But that, as it turned out, would take a long time. One of the most striking—and certainly one of the most intriguing—aspects of the history of ecocentric restoration is how long it took practitioners, having invented it, to *discover* it and to begin to realize its distinctive value as a conservation strategy and a context for negotiating the relationship between humans and the rest of nature. In fact, the launching of this scattering of projects in the 1920s and 1930s proved curiously sterile and was followed by a long period—roughly half a century—during which ecocentric restoration was variously neglected, actively resisted, or simply ignored by land managers, environmentalists, and ecologists. Although conservation was flourishing during this period, few if any saw ecocentric restoration as anything more than a hobby, a novel motif for landscape design, or a boutique form of land management with little or no value for conservation.

This was evident in the fate of pioneering projects such as those we pro-filed in chapter 4. We have seen how the biologists' initiative on behalf of restoration in the national parks failed in the face of competing interests in land management oriented toward human interests. And of the seven early examples of ecocentric restoration we have profiled, all but two were even-tually abandoned, as their champions died or retired and their projects were either absorbed into projects with other objectives or forgotten and allowed to fall into disrepair. At Vassar, Edith Roberts retired in 1948,[1] and the project she had begun fell into neglect. Part of the "out-of-doors" labo-ratory Roberts and her students had created was obliterated by a new building and parking lot in the 1970s, and restoration efforts did not re-sume on the site until the 1990s, when the college undertook a purple loosestrife control program and reintroduced some of the native species that had been lost. Even then there was little long-term institutional com-mitment to the project. In a report written in 2002, after an inspection of the site, landscape architect Dorothy Wurman wrote, "At present, this gar-den has deteriorated to the extent that it is unrecognizable from its original design. The once-celebrated project has been virtually forgotten."[2] In-deed, when a few years before Wurman wrote her report, biology professor Margaret Ronsheim discovered remnants of Roberts's plantings and began taking classes to visit them, she took them to be bits of natural vegetation fortuitously saved because they happened to be in an undeveloped area between two buildings.

Similarly, a thousand miles away at the Cowling Arboretum, after Blake Stewart retired in 1974, European buckthorn and other exotics be-gan to infiltrate the small forest he and Harvey Stork had planted decades earlier. Ambrose Crawford's project at Lumley Park continued into the 1970s, principally because Crawford himself, who died in 1980 at the age of one hundred, continued to lead his Saturday workdays until just a few years before his death. The site was turned back to the Shire Council in 1976 and then languished until the early 1990s, when the Society for Growing Australian Plants took over responsibility for renewing the resto-ration effort. And the project at the Holden Arboretum had long been abandoned in favor of a program emphasizing ornamental horticulture. Brian Parsons recalls discovering the project in the late 1970s when the ar-boretum librarian, whom he happened to be dating, directed his attention to records of the project in the arboretum's archives.

Of the projects we have profiled, just two were exceptions to this pat-tern. One was at the Desert Botanical Laboratory, where the minimalist restorative intervention of fencing has been maintained continuously for more than a century and where in recent years the staff has undertaken additional restoration efforts, notably in the control of exotic buffelgrass (*Pennisetum ciliare*). The other was at the UW–Madison Arboretum, where the restoration efforts, though curtailed somewhat by the departure of the Civilian Conservation Corps, have continued at varying levels of intensity down to the present.

Not Quite

The question we explore in this chapter is, exactly what did these inventors of ecocentric restoration think they were doing? More specifically, to what extent did they come to think of what they were doing as a new form of land management and to recognize its distinctive value, both for the ecosystems being restored and for our relationship with them and with the rest of nature?

The UW–Madison Arboretum project offers an ideal context for exploring this question because it juxtaposed the two parts of this question in a way that was not only unusual but probably unique. Leopold was developing his ideas about the intrinsic value of nature just as he was playing a key role in two projects we now recognize as pioneering ecocentric restoration efforts: one at the arboretum in Madison and one he and his family conducted at their weekend retreat in Sauk County during the same period. The question is whether Leopold, or anyone connected with the project, ever connected the idea with the practice, recognizing that what they were doing was a uniquely powerful tribute to the intrinsic value of nature precisely because it entailed the setting aside, and at times even the sacrifice, of human interests. As far as we can tell, the answer to this question is "no."

Let's consider this in some detail. To begin with, it is clear that Leopold reached the mature version of his idea of the intrinsic value of nature in the decade and a half between the shaping of the restoration plan for the arboretum in the mid-1930s and his death in 1948. Leopold scholar J. Baird Callicott makes this clear when he writes that the idea "that plants and animals, soils and waters are entitled to full citizenship as fellow members of the biotic community, is tantamount to the recognition that they too have intrinsic and not just instrumental value."[3] This idea is fundamental to Leopold's most influential thinking and is the basic theme of his famous essay "The Land Ethic," in which he insists on "the existence of obligations over and above self-interest" with respect to land and its management.[4] Perhaps most directly relevant to the distinction we are making between self-interest and environmental altruism, he noted that responsible behavior with respect to our environment may entail "sacrifice."[5] And in an unfinished and undated essay he argued that "man must assume that the biota has a value in and of itself, separate from its value as human habitat."[6]

For more than half a century, advocates for the environment have drawn on this writing to make the case for the preservation of species and habitats ranging from the snail darter and Furbish lousewort to the Amazonian rainforest. But preservation is one thing; restoration is another, and it arguably offers a more demanding test of one's commitment to the notion of intrinsic value. The question for us is whether Leopold and his collaborators in the UW–Madison Arboretum project realized this and thought of what they were doing as having distinctive value as an expression of, or tribute to, the intrinsic value of the systems they undertook to restore, separate from their value as human habitat.

Here again, Leopold provides our clearest, most reflective account of what he and his colleagues thought they were up to in attempting to "reconstruct" a collection of historic ecosystems. No doubt largely for this reason, it has become common to identify Leopold as the father of ecological restoration—that is, implicitly, ecocentric restoration, because it was the commitment to re-create the whole historic system that distinguished this (and its sister projects elsewhere) from land management practices dating back millennia. Clearly, Leopold played an important role in conceiving this idea and reducing it to practice. More than anyone else doing similar work at the same time, he articulated his developing ideas of what this effort was all about. In 1942, for example, pointing to the ongoing work at the arboretum, then in its ninth year, he recommended a similar project to the staff developing the Agency House Historic Site in Portage, north of Madison, arguing, "We can hardly understand our history without knowing what was here before we were." And in another essay he celebrated the efforts of "two middle-aged farmers" to restore a tamarack bog, a project that offered "no hope of gain" and that he characterized in defiantly nonutilitarian terms as "a revolt against the tedium of a merely economic attitude toward land."[7]

All this makes it clear that Leopold at least, and presumably his collaborators in this and related projects, did see this form of management as a gesture of respect for historic ecosystems, for their own sake. Yet as far as we can tell, neither Leopold nor any of his colleagues ever set this effort apart, clearly distinguishing it from meliorative land management. Strikingly, when Leopold described the project at the dedication of the arboretum in 1934, he justified it with frankly anthropocentric appeals to the utilitarian value of the model associations, specifically in building soil, providing forest products, maintaining various aesthetic assets, and generally "preserving an environment fit to support citizens."[8]

Leopold scholarship clearly supports this interpretation of his take on the pioneering attempts at ecocentric restoration in which he was involved. Leopold famously came to see nature, or what he called "land," as "a community to which we belong."[9] And although he developed an idea of the intrinsic value of nature, argued throughout his career for wilderness and its preservation, and objected to the practice of valuing land exclusively as human habitat, he consistently couched his arguments in terms of human interests. Setting aside not only the utilitarian conservationism in which he had been trained but also the preservationism that has been set off against it throughout the past century, Leopold, Callicott points out, "was primarily concerned, on the ground as well as in theory" not with preservation of ancient or historic ecological communities or landscapes but "with integrating an optimal mix of wildlife — both floral and faunal — with human habitation and economic exploitation of land," a position he sees as going beyond the development versus preservation debate.[10]

Similarly, philosopher Bryan Norton writes that although Leopold "saw new and grave responsibilities limiting human activities" in the industrial era, "whether he saw these obligations as deriving from sources outside of and independent of human affairs seems to me doubtful."[11] And historian and Leopold biographer Curt Meine writes that Leopold downplayed the distinction between self-interest and altruism and between economic and transcendent values such as meaning, beauty, and community. He argues that Leopold moved toward an integration of utilitarian and transcendent values, ultimately taking a position that made it possible to argue for higher values on utilitarian — indeed, human-centered — terms.[12] Meine regards this as an important part of Leopold's achievement. Clearly, identifying altruism with self-interest is an appealing idea, resonating as it does with the myth of the restoration of a "first time" free of existential conflicts of interest. But it leaves out the idea, articulated as we have seen by Leopold himself, that the relationship with a valuable other may entail some sacrifice of self-interest. In fact, it moots the question of self-interest versus altruism by implying that the two are identical. But of course humans, like any other species, influence their environment, and the ecosystems they inhabit are shaped at least in part by their activities. This was spectacularly true in the Midwest, where an array of grassland types had been shaped in large part by human activities, notably firing of vegetation. These were ecological misfits in a landscape occupied by farmers and town dwellers, and they

provided habitat for these newcomers only to the extent that they were willing not only to restore the old ecosystems but then to inhabit them on the
same terms as those who originally inhabited and shaped them.

Of course, that was not at all what Leopold and his colleagues at the arboretum had in mind. They hardly supposed that Madisonians would rein-
habit the arboretum's re-created forests and grasslands as hunter–gatherers
and cultivators of corn, squash, and beans but were interested in the idea
that these ecosystems might serve as models for a healthy ecosystem in
ecological harmony with a human community in a more general sense. To
the extent to which that was true, however, it proved to be true only partly
and in complicated ways. To begin with, the ecosystems Leopold and his
colleagues were attempting to restore showed little sign of behaving as organisms, taking on a life of their own and moving toward some climax condition. Far from offering models of ecological stability, they turned out to
be fragile constructs, clearly defective by ecological and historical criteria
and dependent on continual management to compensate for the altered
environment in which they found themselves. This probably came as no
great surprise. As Callicott points out, Leopold himself had long been
skeptical of Clements's idea of the stable climax community.[13]

Besides that, Leopold's own experience was undermining his confi-
dence in the ability of humans to manage, much less reproduce, any kind
of ecological system—or at least to do so deliberately.[14] Philosopher Eugene Hargrove writes that by 1936, just two years after the arboretum planners had laid out their plans for the restoration effort, Leopold had abandoned the idea that managers could invent artificial substitutes for the
qualities he regarded as characteristic of natural associations.[15] Responding to the failure of single-factor attempts to solve the problem of managing deer populations in the Southwest, Leopold moved toward the position that "such problems probably cannot be solved"[16] and adopted a
management philosophy that Hargrove calls "therapeutic nihilism." In the
nineteenth century, Hargrove notes, physicians, impressed by the complexity revealed by discoveries in physiology and pathology, despaired of
their ability to restore health and began to rely more and more on the self-
healing ability of the body to cure itself of injury and disease. Hargrove argues that Leopold made a similar move in the 1930s as developments in
ecology, including his own work in wildlife management, revealed the
complexity of ecological systems. The idea was that if attempts to manage
a single species by regulating predation often failed, the likelihood of reproducing an entire community must be vanishingly small.

In response, Leopold did not abandon the idea of restoration, but he
changed his idea of its objectives. From the late 1930s on—the last decade

of his life—he was deeply involved in restoration efforts at a number of sites, but these emphasized conservation of soil, water, wildlife habitat, and other resources, efforts that entailed abandoning the idea of restoring or re-creating actual historic ecosystems and attempting instead to invent novel associations of plants and animals that could exist in harmony with their human inhabitants. These would be defined not by concrete attributes such as the species composition and structure of historic models but in terms of abstract qualities such as productivity, integrity, beauty, and—preeminently—health. These attributes raised questions of their own, in part because they are themselves hard to define in operationally robust ways. Partly for this very reason, however, they constituted a larger target for the restorationist to aim for, one that would be easier to hit than climax associations that turn out to be no such thing.

Callicott regards this as an advance that took restoration beyond the initial, naive idea of restoring historic associations and replaced it with a forward-looking response to environmental problems such as deforestation, soil erosion, and loss of habitat for noneconomic species. Fair enough. But this is not the discovery or development of ecocentric restoration. It is meliorative—that is, self-interested—conservation informed by a concern for noneconomic species. It is strongly reminiscent of, if not identical to, the more successful land management practices of traditional people described by Fikret Berkes and his colleagues. It construes the land in edenic terms, as habitat for humans, and although it allows for a generous complement of other species, presumably including "useless" ones, it is fundamentally anthropocentric. "For all his disenchantment," historian Donald Worster writes, "[Leopold] never broke away from the economic view of nature. In many ways his land ethic was merely a more enlightened, long-range prudence . . . [and] he continued to speak in agronomic terms; thus the entire earth became a crop to be harvested, though not one wholly planted or cultivated by man."[17] This leaves open the question of what would become of the old ecosystems championed by Leopold and many others in a landscape managed in this way. It also raises a philosophical question: What does the idea of intrinsic value really mean, what does it demand of us, once we decide that our interests are ultimately identical with those of everything else in our environment?

Weekends

But if Leopold put the idea of ecocentric restoration on a back burner in the last decade of his life, it is important to keep in mind that he never fully lost interest in it. Although he seems to have spent little time at the arboretum after the first few years of the project, and in the last decade of his life he devoted much of his attention to various projects aimed at integrating the interests of humans and nature, he maintained a commitment to the experiments undertaken at the arboretum and undertook a parallel project on a smaller scale at his family's weekend retreat in central Wisconsin. Callicott, having credited Leopold with moving quickly past the ideas underlying these projects to a philosophy of land rehabilitation not unduly concerned with historic models, sees these pursuits as reflecting a merely "quaint" and "antiquarian" interest in the land that is presumably not an important part of Leopold's legacy.[18]

From our perspective, however, this is overlooking an important point. Words such as *quaint* and *antiquarian* are often used dismissively to evoke the image of a lot of useless old stuff in dusty cabinets. But showcasing objects in this way, deliberately and emphatically isolated from contexts in which they are merely useful, is a powerful way of paying tribute to their intrinsic value—that is, their value independent of use.

This is why the earliest experiments in this form of land management were undertaken in privileged situations such as academic and eleemosynary institutions or were pursued as a hobby—that is, as play. If the UW Arboretum was a playground of sorts, then the Leopolds' rural retreat, free of even the slight constraints presented by a land grant university, was even more so. Leopold and his family could work—or play—there without having to justify the effort as research or beautification, as practical or even sensible.[19] So it makes sense that he pursued this odd and impractical form of land management there and on weekends—the secular Sabbath—rather than on Wisconsin farmland, or even at the taxpayers' expense at the arboretum. It is also worth noting that it was this experience that provided the grist for his classic *Sand County Almanac* and that it was in that experience, Leopold reported, that he and his family found their "meat from God."[20]

Still, if Leopold did continue his tinkering with ecocentric restoration in the seclusion and freedom of rural Sauk County, his account of this effort in the *Almanac* suggests that the experience only reinforced his skepticism about the practical, ecological prospects of such an effort. The *Almanac* is explicitly a series of reflections on Leopold's experiences at his family's rural retreat doing, at least in part, what we would call ecocentric restoration.[21] What is striking about these essays, however, is not only their elegiac tone but the downbeat assessment of the prospects for restoration of actual, historic ecosystems they convey. In what is perhaps the most explicit account of such an effort—his attempt to reintroduce to the property silphium, a striking, sunflower-like plant that Leopold casts as an emblem of the tallgrass prairie—the message is one of failure. Having tried unsuccessfully to transplant a specimen of this deep-rooted plant, he tries seeds. These, he reports, "came up promptly, but after five years of waiting the plants are still juvenile, and have not yet borne a flower stalk."[22]

No one, Leopold least of all, would take a single such experience as a clear indication that restoration is impractical. But Leopold is clearly presenting this story as a parable, and, read in this way, it certainly conveys this idea. Looking past the fact that the plants are flourishing to emphasize that they have not yet set seed conveys a negative sense of the prospects for the "fertility" of the restoration effort itself. Consistent with this, Leopold does not mention the several dozens of hectares of prairie already under restoration at the arboretum in Madison, prominently including several species of silphium. In fact, he had already implicitly dismissed the long-term promise of that project by asserting three pages earlier that "what a thousand acres of silphiums looked like when they tickled the bellies of the buffalo is a question never again to be answered, and perhaps not even asked" (49).

Dissonances

What was going on here? Why, with the exception of the UW–Madison Arboretum, did all the projects we have identified as at least approximating the idea of ecocentric restoration wind up being abandoned within a few decades after their inception? Why did Leopold wind up tinkering with it more or less as a hobby, apparently without giving much thought to the role it might play in the conservation of the old ecosystems he valued as wilderness? And why did the early experiments in ecocentric restoration, some of which were well positioned at educational institutions, have little or no influence on conservation thinking and practice over the next four decades?

This is a complex question but also an interesting one because the answers reveal—or at least suggest—a good deal about environmental thinking during this period. For half a century conservation thinking had taken place in the field of tension between the two poles of "conservationism" and "preservationism." Restoration of natural resources was consistent with the first of these. But ecocentric restoration, which mixed active management with the aims of preservation, fit neither. From the perspective of conservationism the notion of ecocentric restoration, with its concern about historic and ecological authenticity and its insistence on devoting attention to species that have no value as resources, appeared wildly impractical. This wasn't serious work. It was expensive, even self-indulgent play.[23] From the perspective of the preservation-oriented environmentalism that emerged a generation later, on the other hand, it seemed not only presumptuous but wrong-headed—at best an illusion, at worst a false promise that could be, and sometimes has been, used to undermine arguments for the preservation of existing natural areas.

On top of all this, at a purely practical level, the trend toward specialization among disciplines and professions generally, and toward analysis and reductionism in ecology in particular, during the years after World War II,[24] certainly discouraged interest in an enterprise that was nothing if not cross-disciplinary as well as applied. This discredited it for a generation of specialists, who dismissed it as a form of dilettantism that had little relevance to serious research or conservation efforts.

Yet another factor was the ambivalence about the past that we discussed in chapter 3. Americans have always havered between an ambiguous interest in the past and outright rejection of it or, perhaps more accurately, between an interest in the historic past and a susceptibility to a mythic past. As we have seen, the idea of autonomous succession to a climax community that characterized Clementsian ecology resonated strongly with the idea of a mythic past outside of history, but the unraveling of this idea, in which Leopold himself participated, deprived restoration of this privileged objective. The old associations, it turned out, were just that: old, even obsolete assemblages of organisms, reflections not of a mythic, organism-like integrity but of the accidents and contingencies of history. Unlike the stable associations of Clementsian ecology, these associations were merely historical, ecologically messy, and self-organizing only in a provisional, ad hoc, catch-as-catch-can sense and so were of value as models for land management in only a general, abstract, and complicated sense.

This loss of faith in the past as a roadmap or repository of ideal models reflected a shift in ideas about the past taking place in American society generally during this period. As David Lowenthal writes, by the beginning of the twentieth century history had lost for Americans its air of destiny, as the playing out of a pattern inherent in human nature.[25] From the perspective offered by Darwin and the emerging discipline of history, everything in the past was unique and was no longer seen as representing an ideal or as embodying some kind of cosmic promise. This included ecological associations as well as villages and empires. And because history was still going on, to try to re-create any of these things was simply to defy the currents and eddies of time and patternless change.

Accompanying this idea was a tendency toward a declensionist idea of history—the idea that history reveals an inexorable decline from the condition of an original Golden Age. Historian Arthur Herman argues that this idea is characteristic of many societies and has been expressed especially clearly in the environmentalism of the past generation.[26] Certainly the elegiac tone of *A Sand County Almanac* reflects this idea, as does the dismissal of the very idea of environmental restoration implicit in the phrase "fragile, irreplaceable ecosystems," which was a commonplace of environmental rhetoric in the 1960s and 1970s. Consistent with this, literary critic Terrell Dixon notes that American writing since the publication of Rachel Carson's landmark book *Silent Spring*, itself a jeremiad portending environmental disaster and couched in apocalyptic terms, has been far more concerned with—and adept at—documenting environmental decline than with articulating a vision of a future based on restoration and renewal.[27]

All this took a lot of the fun out of the idea of restoration. It deprived the restorationist of a privileged objective: the "original" prairie or climax forest. It meant that the attempt to restore the old ecosystems would mean undertaking the task of reversing history, which is not only "laborious," as Leopold had said, but flatly impossible. It also meant that the effort would not lead to a homecoming or a return to a condition of primal innocence, health, and integrity. If, as Callicott argues, Leopold had chosen the old ecosystems as models for the restoration efforts at the arboretum not because they were old or historic but because he supposed them to be "a base datum of normality, a picture of how healthy land maintains itself as an organism,"[28] it obviously made sense for him to abandon or downplay historic models in favor of the more practical and more creative pursuit of models of health and integrity adapted to present-day conditions.

The National Park Service, Again

As far as we know, with only scattered exceptions, no one did ecocentric restoration from the late 1940s until well into the 1970s, and those who did were independent types who undertook projects such as restoration of a prairie in Illinois or bird habitat in Bermuda[29] more as a fascinating, perhaps slightly eccentric avocation than as the prototype of a new form of land management that might have important implications for conservation. This combination of neglect and, on the few occasions when the issue came up at all, outright resistance is clearly illustrated by its treatment by the National Park Service (NPS) and The Nature Conservancy.

The NPS offers an important example of how this played out in the context of a large government agency. Having sidelined the biologists and relieved itself of the science-oriented culture that had grown up around George Wright's group in the early 1930s, the NPS entered a period of roughly a quarter of a century during which concerns about restoration, or even ecologically informed management of any kind, were decisively on the back burner, upstaged almost entirely by the NPS's concern for what amounted to box office.

Following the story of land management in the national parks into the postwar years, Richard Sellars describes a period during which the dominant, visitor-oriented culture asserted itself virtually unchecked by the protocols and constraints of science-based management. The NPS in varying degrees half-heartedly embraced, actively resisted, or simply ignored the ethos of research-based management, including restoration.[30] What Sellars describes is basically a continuation of the policy of resistance, not only to ecocentric restoration but to science-based land management generally.

"After removal of the wildlife biologists to the Fish and Wildlife Service in 1940," Sellars writes, "nearly a quarter of a century would pass before any meaningful attempt to revitalize the National Park Service's biological science programs" (149). Indeed, the superintendent at Yellowstone rejected a proposal by Aldo Leopold for a study of the elk herd in the park in 1943 on the grounds that it would be a waste of money and that what was needed was an "authoritative statement" defining policy for management of the herd. This of course made no sense at all and marks the beginning of a long period during which the NPS basically tolerated research while keeping it at arm's length and remained ambivalent about the management of park ecosystems in general.

There were a number of reasons for this. One was the agency's longstanding commitment to visitor satisfaction and the culture that had grown up around it. This had profound philosophical and cultural as well as organizational implications. What was at stake here was precisely the distinction we are making between ecocentric restoration—restoration of complete ecological systems, even when this entails some setting aside of preferences or sacrifice of our own interests—and management of resources for our own use or satisfaction. To blur this distinction, or dismiss it because it poses a philosophical enigma, as Marcus Hall has done, is to miss this point entirely and to offer a distorted account of the history of land management as it played out in organizations such as the NPS. Indeed, as Sellars writes, "Much of National Park Service history since 1963 may be viewed as a continuing struggle by scientists and others in the environmental movement to change the direction of national park management, particularly as it affects natural resources" (217). Here, two cultures were at odds with each other over an issue that was ultimately philosophical and even religious, the issue commonly represented by the debate between John Muir and Gifford Pinchot: Does nature exist solely or primarily for human benefit, or does it have value in and of itself? The NPS, which has roots in tourism and railroad interests, as well as in disinterested devotion to nature, had been in large part Pinchotian from the beginning, construing its mission to be the provision of aesthetic and recreational benefits to the public, and this had changed little over the years.

Besides this, the NPS entered the postwar era with facilities inadequate to meet the demands of what proved to be an increasing number of visitors, who began arriving at the parks in unprecedented numbers. The agency's response was an increasing emphasis on the development of visitor facilities and services, culminating in "Mission 66," a program (named to mark the agency's fiftieth anniversary in 1966) that emphasized public use and, Sellars writes, "was the antithesis of the scientific approach to park management" (173).

Management was carried out in the parks during this period, however, and some of it was restorative. In the 1930s, in response to the urging of NPS biologists, the agency had inaugurated a program of shooting to reduce populations of elk and other ungulates, which had been expanding in some of the western parks for reasons that were not clearly understood. Initially this herd reduction was carried out by NPS personnel, but by the 1950s hunters were pressing for public involvement. This led to controversy, and it was partly in response to this that interior secretary Stewart Udall asked the National Academy of Sciences to undertake a review of the "natural history research needs and opportunities" in the national parks (200). Chaired by Aldo Leopold's son, Starker Leopold, the "blue ribbon" committee set up to carry out this review was in a way a reprise of the team of biologists working with George Wright who had urged a restorative approach to park management a generation earlier. The difference was that, for the first time, the resulting recommendations would come not from a small, beleaguered team of insiders but from outsiders charged specifically with taking a critical look at management policies and practices in the parks—something that, Sellars notes, had never happened before. This marked, he writes, "a new era, in which park management would be judged far more on ecological criteria" (203).

But not right away. The resulting report, which came out in 1963 and came to be known as the Leopold Report, reads today like a straightforward prescription for a program of ecocentric restoration in the national parks. Indeed, it is almost as if Starker, who graduated from the UW–Madison in 1935, had picked up the ideas his father and his colleagues had developed at the UW–Madison Arboretum and projected them onto the parks. Not only did the report assert ecological criteria for management, it was also history-conscious, arguing that "the objective of ecologic planning in the parks" should be to maintain parks as vignettes "of primitive America," offering "a natural scene that was observed by the pioneers . . . or whoever was the first visitor to the area." In a similar spirit, a plan the NPS devised a few years later outlined a comprehensive management program that took into account the vegetation as well as the wildlife that had been the focus of earlier management policy and debate (246). Setting aside the notion that this could be achieved simply by protecting the parks from outside influences, the committee noted that such influences were inevitable and that it would be necessary to compensate for them in order to keep the parks in their precontact condition. And, while urging reliance on natural processes as much as possible, the report also recommended measures such as prescribed burns as the most "natural" (254) way of doing this.[31]

This was the first major assault by biologists on NPS policy in a generation, and, Sellars writes, the agency reacted defensively, resisting the report's recommendations and looking for ways to downplay it within the agency and to minimize its exposure to the public (214ff). The reasons for this had less to do with fundamental unease in response to the notion of managing for natural qualities—everyone approves of "nature" in the national parks—than with the NPS's long-standing commitment to visitor-oriented management and, complementing that, resistance to the idea of managing to ecological criteria. Although the report urged the agency "to 'restore and maintain the natural biotic communities' in the parks, the acting assistant director for the agency's forest and wildlife rangers wrote in 1966 that these communities have 'little justification for retention as national parks except as they are utilized by man, i.e., the park visitor.'" Reflecting on such attitudes, NPS scientist "Lowell Sumner asserted that the 'trouble with ecological considerations' in the parks had been that they were 'frequently in conflict with some of the programs of other Service units—programs such as native forest insect control, filling in of swamplands to enlarge campgrounds, road and trail building into essentially pristine ecological territory, or suppression of natural fires in parks whose distinctive vegetation was dependent on the continuing role of natural fires.'"[32] What lay at "the heart of the matter," Sellars writes, was what Sumner described in 1968 as the NPS's "reluctance to acknowledge the ecological importance of the parks."[33]

The outcome was not a half but perhaps a quarter of a loaf—more accurately, a loaf deferred. Sellars characterizes the Leopold Report as an important watershed in the history of NPS management policy, after which a slow buildup of scientific expertise began in the agency and "scientific and ecological factors became the chief criteria by which the Park Service's natural resource management—and much of its overall management—has since been judged" (267). At the same time, characterizing the agency's overall response to this development and the ideas behind it as "sporadic and inconsistent," Sellars finds that little progress was actually made on the ground at the time and that the promises associated with this new dispensation were "largely rhetorical" (269). Indeed, as though acting out a thirty-year cycle that had begun in the 1930s with the efforts of George Wright and his colleagues, a series of conferences held in the early 1990s also urged greater emphasis on research-based, restorative management on lands such as those in the national parks, but again with little immediate result.

The Nature Conservancy

Turning to The Nature Conservancy (TNC) as a case study in the response of a nonprofit environmental organization to the prospect of ecocentric restoration, we find a roughly parallel pattern of neglect through the 1950s and 1960s. TNC was founded in 1951 to preserve natural areas. However, it traces its origins back to 1917, when members of the activist wing of the Ecological Society of America, formed three years earlier, created the Committee for the Preservation of Natural Conditions. In 1926 the committee, chaired by ecologist Victor Shelford, published *The Naturalist's Guide to the Americas*, an attempt to catalog all the known remnants of wilderness left in North, Central, and South America, Bermuda, the Antilles, the Galápagos, and the Philippines.[34]

This proved a landmark in the history of attempts to ensure the survival and well-being of historic ecosystems. Ecologist Robert Jenkins, who a half a century later launched TNC's first attempt to organize both the acquisition and the management of TNC lands on an ecological basis, recently called it the opening shot in the development of the inventory plans that would eventually become the foundation and hallmark of TNC's program.

Like the National Park Service, TNC had a mission that committed it to work for the long-term survival and well-being of natural, or what we are calling historic, ecological systems. But the organization did this at first mainly by acquiring land and paid little attention to the question of how to manage it. When Jenkins joined the staff in 1970 he found that the only paper in the office that pertained to management of TNC's holdings was an all-but-forgotten Policy Bulletin No. 2, compiled some years earlier by Daniel Smiley, an early board member, which provided some guidelines based, Jenkins suspects, on Smiley's experience in managing his own extensive properties. The other bulletins in what was presumably a series were lost, and Jenkins soon realized that this reflected TNC's priorities at the time.

"Our motto was Land preservation through private action," he recalls. "But we had no idea what preservation meant. What it meant in practice was grab anything you can get your hands on. It wasn't clear whether the aims were ecological quality, historic preservation or even maintaining open space for recreation. There was no systematic way of prioritizing acquisitions, and the question of management philosophy was irrelevant. There was none. The Conservancy had a staff of 30 or 40 at the time, and not a single one responsible for, or trained to look after, the management of our holdings. In fact, we hardly knew what we had. One of the first things I did after joining the staff was to spend some time driving around to take a look at our sites in the D.C. area, and I soon found that I couldn't even find some of them. Most were overseen by committees made up of people who had spearheaded the campaign for acquisition and had little or no ecological background. In one case I found that the guy in charge of an area had been dead for two years."

Backing up Jenkins's on-the-scene recollections, when philosopher Anthony Smith scanned issues of *Nature Conservancy Magazine* from its first issue in 1951 down to 2006, he found only incidental references to restoration before 1970. This may have reflected a conventional wariness about tinkering with nature, but Jenkins believes it had more to do with the land acquisition–oriented and business-oriented culture of TNC in its early years—actually one reason for its survival and long-term effectiveness. He notes that once presented with proposals for specific management projects, the organization offered little resistance and, in fact, sponsored a number of projects in different parts of the United States within a few years of Jenkins's arrival. Given TNC's mission, these were basically ecocentric restoration projects, and although it did not simply embrace restoration overnight, TNC was to prove much faster on its feet in this matter than the NPS.

Irreplaceable Ecosystems

TNC's neglect of restoration in the 1950s and 1960s was typical. As the environmental movement got under way during this period, it fostered a rhetoric of preservation, which cast natural ecosystems not only as fragile but also as irreplaceable, reflecting a philosophy of nature and management that precluded serious consideration of restoration. The extent to which discussions of environmental conservation were dominated by this rhetoric of preservation, even late in this period, was brought home to Bill Jordan at the symposium on biodiversity organized by the National Academy of Sciences and the Smithsonian Institute and held in Washington, D.C. in 1986. The three-day symposium included a session on the role restoration might play in conserving biodiversity. It wound up with a special panel in which conservation leaders Paul Ehrlich, E. O. Wilson, Peter Raven, Tom Lovejoy, Michael Robinson, and Joan Martin Brown exchanged remarks summing up their impressions of the presentations of the previous two days. Jordan went to this discussion wondering what these prominent observers would say about the session on restoration, in part because he had helped organize it but also because restoration was at the time something of a novelty in discussions of this kind. Within a quarter of an hour, however, he realized that they wouldn't have much to say about it, not because they considered it irrelevant but because the conversation was couched in the rhetoric of preservation, so that bringing up restoration would have come across as not only changing the subject but as contradicting the main message. The discussion, which ran nearly two hours and included remarks by Wilson that conveyed a more optimistic perspective than had prevailed through most of the symposium, concluded without a single reference to restoration.[35]

This illustrates what was a real problem for advocates of biodiversity or natural ecosystems generally. To make the case for preservation of natural ecosystems, they naturally had to make a case for their protection from outright exploitation or conversion to other uses. In practice, this usually meant talking the language of preservation, which cast ecosystems such as ancient grasslands, old-growth forests, or coral reefs as having a unique value precisely because they are natural—that is, owe nothing to human (or, more accurately, Western or modern) influence. This meant stressing the intrinsic value of these ecosystems, a value that, at least by some readings, could only be compromised by even the best-intentioned human tinkering, including tinkering aimed specifically at compensating for human influence.

This made restoration a hard sell for a public schooled on decades of preservationist rhetoric and perhaps troubled by what biologist Walter Rosen, who had represented the National Academy in organizing the symposium on biodiversity, once called the "loss of innocence" inherent in the practice of restoration.

Compromised by the human touch, a restored "natural area" was from this perspective an oxymoron, inherently inferior to its natural counterpart. And this being the case, restoration was best viewed as an emergency response to environmental harm—a response that characteristically cast the restorationist as the good guy, rather like the doctors in the old TV series *M.A.S.H.*, who take no responsibility for the messes made by the bad guys—the war makers, developers, or surface miners—but only come in as high-minded heroes to clean them up.

At a more practical level, environmentalists were naturally concerned that the promise of restoration could be used to undermine arguments for preservation. For example, Eric Higgs notes that prominent environmental leader David Brower initially "rejected" restoration when it was gaining the attention of conservationists in the 1980s, arguing that it "should be opposed at all costs" because it would distract environmentalists from the essential task of preservation.[36] Others recognized the value of restoration but worried about misuse of the idea. Michael Fischer, who was then the executive director of the Sierra Club, argued at the "Restore the Earth" conference in Berkeley in 1988 that restoration might have an important role to play in some situations but must never be accepted as a form of "mitigation" to justify harm rather than reverse it. When we asked Fischer about this recently, he recalled the comment and emphasized the "stern, quite negative tone that I adopted. A lot of the people there were consultants for developers seeking justification for mitigation banks and restoration projects to enable proposed wetland destruction projects. And, yes, I criticized them for what I felt subverted the legitimate business of restoration ecology."[37]

Similarly, ecologist Joy Zedler began publishing papers voicing her concerns about wetland restoration in the 1980s, cautioning against an overly sanguine view of restoration as the solution to the problem of wetland destruction. On the basis of exhaustive assessments of processes such as nutrient cycling and productivity as well as detailed inventories of species present, Zedler and her colleagues pointed out that the restored tidal wetlands they investigated near San Diego differed markedly from the natural wetlands they were intended to replace.[38] Here again, like Fischer, Zedler was not opposed to restoration (indeed, since 1998 she has been the Aldo Leopold Professor of Restoration Ecology at the UW–Madison), only to the uncritical acceptance of it as an alternative to preservation. This was a concern that had a special urgency at a time when some were proposing just that in response to the policy of "no net loss" of wetlands then being promoted by President George H. W. Bush. But the risk on the other side was that by downplaying the value of intensive restoration as a way of *replacing* wetlands destroyed outright to make way for the construction of highways and shopping malls, conservationists might lose sight of the role what we might call chronic or ongoing restoration has to play in the conservation of *all* old ecosystems subject to novel influences—that is, ultimately, all of them.

Managers usually refer to this sort of low-key restoration as "stewardship" or "management," soft-edged words that sidestep both the promise of restoration and the many questions it raises. Bill Jordan recalls an exchange at the 1990 colloquium that resulted in the edited volume *Beyond Preservation*.[39] When a member of the audience asked ecologist Orie Loucks whether the distinction between restoration and long-term management aimed at "preservation" isn't a false distinction, Loucks replied that it clearly is but that "people might not be ready to hear that"—in other words, to hear that "preservation" *depends* on restoration and that an ongoing program of restoration is called for whenever the objective is the long-term survival and well-being of a historic ecosystem. In the end, the dichotomy remained a persistent theme of the conference. Some of the participants were severely critical of the whole notion of restoration, seeing it as yet another expression of human arrogance in its manipulation of nature, as an exercise in nostalgia, or as a self-satisfied conceit indulged by a privileged, yuppie class.

Others, including Loucks, acknowledged that despite its limitations, restoration may be called for in some situations, but they had qualms about embracing it as the best—if imperfect—way to ensure the perpetuation of ecosystems and the species that inhabit them. Loucks, for example, provided downbeat accounts of restoration efforts in Switzerland and China, contrasting them with the story of a preserve in southeastern China. There, he noted, a fifth of the species now present are "introductions." Yet he celebrated the preserve as "a scientific and spiritual wonder" because it had been unmanaged, or managed only minimally over the years, and the accessions (and presumably losses) of species had been un-intentional—"a feat of nature that no one can contemplate as a restoration."[40]

All this, as we suggested at the outset, reveals at least as much about the character of environmental thinking during this period as about the nature and limitations of restoration. What is important for us is what it suggests about how a wide range of thinkers and observers with a serious interest in the environment and in environmental conservation responded to the challenge inherent in the idea of ecocentric restoration—the realization that this form of land management is not an alternative to preservation but the necessary means of achieving it.

As we are seeing, there has been much resistance to this idea. When biologist Jared Diamond asked in a 1992 article, "Is It Necessary to Shoot Deer to Save Nature?," he concluded that managing nature preserves to maintain their "status quo," a task that often entails measures such as shooting deer, though an "odious task," is ultimately unavoidable if the aim is to maintain the biodiversity they represent.[41]

It is worth pointing out that the ambivalence managers and environmentalists have often expressed with regard to the prospect of ecocentric restoration strongly underscores the salience of the distinction we are making between ecocentric and resource-oriented conservation. Although the commitments of individuals and societies to resource conservation and the various forms of meliorative restoration naturally vary widely, few if any have ever objected to the notion of resource conservation in principle. Ecocentric restoration is a different matter. Calling in many cases for the setting aside of human interests and presuming to recapitulate the work of creation itself, it naturally provokes objections that meliorative forms of land management do not.

In the previous chapter we mentioned that one reason why the restoration effort undertaken at the UW–Madison Arboretum has gained recognition as a landmark in the development of ecocentric restoration was that the arboretum project survived its founders and, despite the compromises the restorationist always has to make with time and circumstance, remained faithful to their commitment to the restoration of whole ecosystems. This was true in the decades after the war and remains true down to the present, a fact that is underscored by the head-scratching that managers at the arboretum are currently doing as they contemplate the challenge of pursuing—or modifying—their mission in response to developments such as increasing runoff from urban development around the property and the early signs of global climate change.[42]

Even at the arboretum, however, the restoration effort has varied in intensity over the years. Routine restorative management, such as burning of the prairies, has continued uninterrupted over the years. But the restoration effort generally declined after John Curtis's death in 1961. Reflecting the concern about toxic chemicals aroused by Rachel Carson's *Silent Spring*, which was published the following year, the Arboretum Committee limited the use of herbicides on the property, lowering the arboretum's defenses against invasive exotics such as Eurasian honeysuckle and buckthorn, which proliferated rapidly in many of the restored communities. They also backed off on thinning of trees as an adjunct to restoration efforts. As a result red and white pines, which had been overplanted decades earlier in anticipation of losses in a forest creation project, had developed into virtual dog-hair thickets dominated by spindly, thirty-year-old trees shade-pruned to mere toothbrushes of canopy foliage. Reflecting the emphasis on preservation that characterized the environmentalism of that period, researchers began to treat the arboretum as a preserve rather than a collection of ecosystems undergoing restoration. Tellingly, in 1970, when arboretum managing director Roger Anderson and UW botany professor and Arboretum Committee chair Grant Cottam prepared a proposal to the National Science Foundation for creation of an environmental research facility at the arboretum to complement its outdoor resources, they emphasized its accessibility, ecological diversity, and well-documented history as key assets but mentioned the role of restoration in the creation of this diversity only in passing. In fact, they referred to the arboretum's restored areas collectively as "natural communities,"[43] and this characterization, which ignored, and indeed implicitly falsified, their history, was typical. Bill Jordan, who began working at the arboretum in 1977, recalls the director regularly referring to the property as a "preserve." Working with historic photographs, he also noticed that people, well represented in the superb, large-format black-and-white photos taken at the arboretum during the Civilian Conservation Corps era, rarely appeared in photos taken in more recent decades, an omission he read as an unconscious reflection of the human-wary preservationism that characterized much environmental thinking during this period. Research continued, but much of it was unrelated to the restoration effort or was descriptive research on various aspects of the ecology of the partly restored ecosystems. Even at the arboretum, a landmark, if not a Kitty Hawk, in the history of ecocentric restoration, interest in restoration of historic ecological communities had reached a low ebb.

Chapter 6

Realization I: Stepping-Stones

What some have called a "movement" eventually grew up around practices first tried out in places such as Broken Hill, Lumley Park, and the Holden, Cowling, and UW–Madison arboreta, but that was a long way off at midcentury. Certainly, resource conservation efforts pioneered in the half century since the era of Gifford Pinchot and Teddy Roosevelt continued, promoted by agencies such as the Soil Conservation Service, Fish and Wildlife Service, the US Forest Service, and an array of state conservation departments, all supported by emerging professions such as forestry and game management. But management was understood to be about resources—soil, water, timber, scenery, and fish and game. Preservation, which gained prominence as a concern among environmentalists in the 1960s, was another matter. Farm fields and woodlots, rangelands, public hunting grounds, and even fishable lakes might be managed for maximum yield of one resource or another. Land set aside as preserves in places such as national parks or on holdings of organizations such as The Nature Conservancy were understood to be just that: preserves. They were understood to be primal, even "original," ecologically pristine, or undisturbed—relics of an edenic past, models of ecological health and repositories of a kind of existential purity that could only be compromised by even the best-intentioned management, including attempts at restoration.

What we find looking back is a long period in which restoration—that is, the version of restoration we are exploring—was generally ignored, marked nevertheless by scattered projects that, increasing in number over the years, eventually catalyzed a new awareness and wider acceptance, and even what some have called a revolution in environmental thinking and conservation practice. In the tallgrass prairie region, which became a prominent incubator and proving ground for ecocentric restoration and have provided examples of the craft of restoration in its most ambitious form, what we find during this period are isolated examples, all of them interesting in various ways but unconnected with what was going on in conservation generally.

In 1942, for example, University of Illinois ecologist Victor Shelford undertook restoration of prairie on 8 hectares of abandoned cropland that the university had acquired a few kilometers from its Urbana campus. Steve Buck, who is manager and research technologist for the university's Committee on Natural Areas, notes that the project, intended to add to the existing Trelease Forest an ecosystem that had been virtually eliminated from the landscape, reflected Shelford's interest in the "humanizing" of ecology, a predilection that put him somewhat at odds with the Ecological Society of America.[1] The project has been maintained down to the present and has been the site of research on fire ecology, the biology of voles, and, currently, the response of ecosystems to climate change.[2]

Green Oaks

Just three years later, in 1945, Henry Greene began his prairie restoration project at Madison, and then there was a lull of an entire decade, during which Americans concluded a war, embarked on a cold war, moved to the suburbs, and went shopping before we picked up another project. This was at Green Oaks, a tract of 285 hectares of land that Knox College began acquiring piecemeal in the countryside not far from its campus in the northwestern Illinois town of Galesburg in 1955. As had been the case at the UW–Madison two decades earlier, the moving force for a restoration project at the site was a handful of faculty and conservation-minded citizens, notably zoologist Paul Shepard, then a young faculty member at the college, who later gained a reputation for his writings on traditional cultures and human relations with nature. Shepard first visited the site with several colleagues in the fall of 1954, and shortly after that the group paid a visit to the UW–Madison Arboretum. They returned full of enthusiasm for the prospect of restoring prairie on the site and an offer from Greene to provide seed for the startup effort. They made their first planting the next spring, covering roughly 5 hectares, and over the next decade their restoration efforts included planting pines and reintroducing native plant species to several aquatic ecosystems.[3]

What they had in mind is recorded in memos, including a brief, unpublished memoir of the project Shepard wrote years later, which make clear their interest in the restoration of the prairies, forests, wetlands, lakes, and streams that had existed in the area—indeed on the site—at the time of European settlement. "What needed to be done," Shepard wrote in his memoir, "was a mammoth, long-term job of restoration, protection and education."[4] By *restoration* he clearly meant re-creation of actual ecosystems and landscapes characteristic of the period of European settlement, and biological authenticity was clearly a priority. In fact, Shepard apologizes at one point in his memoir for using seed provided by colleagues in Madison, some 230 kilometers north of Galesburg, noting that "taxonomically it may have been a mistake to use the Wisconsin material, but we needed prairie vegetation to signal our beginning, even at the cost of biogeographical errors."

But the Green Oaks group was as concerned with history as with ecology. In an interview years later Shepard noted, "We were not only planting prairie, we were trying to build a log cabin of the type that had been built along the prairie edges" during the early stages of European settlement.[5] "The thing that I wanted those students to do," Shepard wrote, "was to stand in the middle of this thing and at least get some sense of what it was like to be in it, to have to deal with grasses four feet taller than you were, how you go about coping with this place as a settlement place."[6]

This emphasis on restoration as a way of reliving history and also on participation by the public and the development of a broad constituency for the project was characteristic of the work at Green Oaks and contrasted with the earlier work at Madison. There the project planners, acting on the idea that the original ecosystems were models of land health that had been compromised by post-settlement history, had pursued what amounted to an attempt to obliterate that history and made a point of tearing down the old buildings on the site, including several that dated back to the late nineteenth century and might have been regarded as valuable emblems of the settlement era. The arboretum managers were wary of contemporary humans as well, regarding them not as a potential asset but as an unavoidable nuisance. In contrast, the work at Green Oaks reflected a much friendlier attitude toward the public and, like Ambrose Crawford's project at Lumley Park, to some extent anticipated the development of restoration as a volunteer-driven, community-oriented activity that practitioners in the Chicago area would pioneer decades later.

Perhaps reflecting the influence of Leopold's later writings but explicitly rejecting the appeals to utilitarian benefits Leopold had relied on in making the case for the project in Madison two decades earlier, the Green Oaks planners expressed their aims almost entirely in terms of the intrinsic value of the ecosystems and landscapes they aimed to restore. Worrying that some might find the project as they conceived it "too idealistic" and foreshadowing the preservation-oriented environmentalism that emerged a decade later, they insisted that there is "a valuable aspect to land which does not depend upon its commercial value" or its value as "a great storehouse of resources to be made into human wealth."[7] As we have seen, Leopold had articulated this idea quite clearly. But as far as we know, he never argued for the restoration effort at the UW–Madison Arboretum in these terms or characterized this kind of land management as a distinctive response to or expression of that idea. In fact, this would have been inconsistent with the integration of utilitarian and higher values that scholars such as Curt Meine and Baird Callicott regard as an important feature of Leopold's mature thinking. In contrast, Shepard reflected in his memoir that the message implicit in the project "has something to do with our recognition that there is an element essential to life which eludes us when we try to shape it to our own ends, and which demands recognizing and valuing the natural world for its own sake."[8] Like George Wright's contingent a generation earlier, the Green Oaks managers found themselves in opposition to the "designers," in this case Alvah Green, who at the time owned most of the property. Green constantly tinkered with the land in order to enhance various aesthetic elements, at one point even bulldozing vegetation in order to improve the view of a favorite display of flowering vines. In his memoir, Shepard devotes much space to deploring what he saw as Green's ecologically uninformed shenanigans.

At the same time, their rhetoric did include, at least implicitly, an appeal to utilitarian benefits in the character of the model associations as they conceived them. Like Leopold and his colleagues, they cast these as models of healthy ecologies, which they characterized in terms of ecological balance and equilibrium, offering a better model for human habitat—aesthetically and psychologically as well as ecologically—than ecosystems that had been disrupted by land uses introduced by European settlers during the nineteenth century.[9]

These documents also reflect a good deal of ambivalence about the whole notion of management and, at least implicitly, the question of whether the aim of the project was to re-create a self-organizing wilderness or a collection of historic ecosystems and landscapes. On one hand, for example, Shepard eagerly embraced restorative measures such as seeding and burning hectares of prairie and reintroducing native plants and even turtles into a lake on the property. On the other, he expressed deep skepticism about management or manipulation of ecosystems generally. At one point he recalls that May Theilgaard Watts, a naturalist at the nearby Morton Arboretum, told him that he should stop fretting about the torn-up land left in the aftermath of coal mining operations that had been carried out on some 80 hectares of the property, insisting that "the worse the strip-mining practices the better the wilderness it would eventually produce." "And she was right," he comments. "The 200 acres of strip-mined land at Green Oaks is becoming a wilderness." "Its lovely lake," he writes, commenting on the large artificial lake the mining operation had created on the site, "is probably the most singularly attractive feature at Green Oaks." And he goes on to caution against the temptation "to do something to it or about it" in an attempt to improve it.[10] As for the mining operation itself, he argues in the proposal that "this extreme violence" is "superlative in its own way," and he celebrates the resulting landscape as "an extraordinary panorama of built hills lying in nearly geometric patterns." "It is," he concludes, "a raw and fascinating area." He notes its value as a case study in ecosystem recovery, then undercuts his admonition to leave the area alone by adding, "With little effort it is possible to guide its slow recovery in any of a thousand directions and forms."[11]

These comments are of great interest for us because they suggest that although Shepard devoted much effort to attempts to restore historic ecosystems on the site, he entertained contradictory ideas about which alterations in the land constituted damage calling for restoration and which were desirable. Plowing down prairie to plant crops was bad, but strip-mining hundreds of hectares and leaving them turned upside down resulted in an ecosystem that could be described as "superlative" and that had a value that would only be compromised by restoration efforts. What this suggests is that, at least for Shepard, the real aim was not the restoration of historic associations, landscapes, or ecosystems but the recovery of wilderness—that is, the condition of being free of human influence even if, as in the case of the mined land on the Green Oaks site, the result would be ecologically indeterminate at best.

Quite apart from its interest as one of the few ecocentric restoration projects undertaken in the third quarter of the twentieth century, the Green Oaks project is of special interest to us for two reasons. One is that it survived the departure of its primary instigator. Shepard left Knox College in 1963, and two years later zoologist Peter Schramm took over as supervisor of the work at Green Oaks, a task he has pursued energetically down to the present. Besides this, Schramm linked up with others in the region as they initiated restoration projects in the 1960s and 1970s and participated in the planning for a North American Prairie conference at the college in 1968. That turned out to be the first of an ongoing series of biennial conferences that has promoted restoration in state parks and along highway rights-of-way, contributing to the emergence of a restoration culture—a movement of sorts—in the region. This had not happened before, and it was an important step in the discovery and realization of ecocentric restoration that we are exploring here.[12]

The second reason why the early work at Green Oaks is of special interest to us is that Shepard makes an interesting comparison with Leopold, who had played a similar role at the UW–Madison Arboretum two decades earlier. This is especially true because Shepard, like Leopold, left behind a large body of reflective writing, much of which, though not about Green Oaks or even restoration, is relevant in many ways to both. In this writing Shepard explores ideas of nature and time, values, and human relations with animals that are only hinted at in the early records of the Green Oaks project. Especially pertinent for us, in his book *The Tender Carnivore and the Sacred Game*, which was published a decade after he left Knox College and became something of a cult object among some environmentalists, Shepard pursued the idea of the "Paleolithic" past as a kind of ideal that is evident in only a sketchy way in the Green Oaks documents. This contrasts with Leopold's more pragmatic, forward-looking thinking about land use. Similarly, projecting this idea into the future, he sketched out a vision radically different from the one Leopold articulated in the later years of his life. Leopold had moved past the version of restoration he and his colleagues had experimented with at the UW–Madison Arboretum to a pastoral vision based on the reconciliation of human interests and the well-being of ecosystems. The result was a deeply appealing idea of human relations with their environment, but it left open the question of what was to become of the old ecosystems in such a "reconciled"—which is to say domesticated—landscape. In contrast, Shepard insisted that "nature and culture need to be separated in the human economic sphere" in order to preserve the experience of nature as other—an important idea for us because it is precisely the commitment to restore ecosystems experienced as *independent of* us that distinguishes ecocentric restoration from other forms of land management. To illustrate this separation, Shepard outlined a vision of a future for North America in which human populations would be concentrated along the coasts, leaving vast areas—indeed much of the interior of the continent—largely uninhabited. The existing ecosystems of this vast area presumably would not be restored but, like the mine spoils at Green Oaks, would be left alone, and the resulting wilderness would provide a context for ritualized traditional activities oriented around the economies of hunting and gathering, serving humans mainly as a context

for identity-forming experiences for the young.[13] Farming in the traditional sense, entailing a collaboration with nature at the ecosystem level in
farm fields, gardens, pastures, and fish farms, would be replaced by wholly
artificial, spaceship-like systems in which microorganisms produce food
and fiber from raw materials such as sewage and other forms of organic
waste.[14]

Shepard proposed this radical vision as a challenge to what he saw as a
civilization based on a "crisis" in its relationship with nature that began
with the development of agriculture. It is of great interest to us because,
taken together, Shepard and Leopold represent two sharply contrasting responses to the ambiguities encountered in the attempt to manipulate an
ecosystem "for its own sake" that defines the practice of ecocentric restoration. Leopold's response to this ambiguity was his vision of a reconciliation
between nature and culture embodied in working landscapes that would
accommodate a diversity of species and processes. And although Leopold
continued to advocate for wilderness and its preservation, he apparently
never recognized the crucial role ecocentric restoration, as exemplified by
the arboretum experiment, had to play in the preservation of actual ecosystems. His vision of a landscape in perfect harmony with itself and also
with its human inhabitants left room for wilderness but implicitly abandoned the old ecosystems to drift ecologically in response to changing
conditions.

In contrast, Shepard put wilderness up front, calling for the near abandonment of vast areas precisely in order to reconstitute wilderness (though,
again, not necessarily actual ecosystems) and to segregate it from the means
of subsistence.

The contrast is revealing. Whereas Leopold moved toward a people-
friendly land management paradigm that some have characterized as utilitarian or anthropocentric (though now making room for economically
"useless" species), Shepard moved toward something like Deep Ecology,
with its biocentric claims that would cede vast areas to other species. And
although both placed a high value on wilderness, it seems that neither recognized that the "sandbox" experiments in ecocentric restoration that they
had helped launch in Madison and Galesburg provided the paradigm for
the preservation of actual ecosystems.

Coalescence

As we have seen, ecocentric restoration, which had been invented more or
less independently by a scattering of practitioners in the 1930s, had faded
from sight as these projects were abandoned in the decade and a half after

about 1940. What we are seeing here, in projects such as the one at Knox College, is a new generation of tinkerers who picked up on earlier projects in various ways but still worked pretty much on their own in response to some abuse or alteration of an ecosystem that for one reason or another they perceived as a challenge rather than an irreversible calamity.

However, these were isolated projects, sparks of an idea few if any recognized as a distinctive — or even useful — conservation strategy, much less an important topic of conversation, a major challenge, an intriguing opportunity, or the basis for a land management protocol or discipline. In a crucial, practical sense, these projects didn't matter, and they wouldn't matter until those involved became aware that they were doing something distinctive and began to exploit this work for its distinctive value in a self-conscious way.

Ecocentric restoration could not really happen until ecologists began to take it seriously — and until practitioners began to take ecology seriously. John Cairns Jr., who has done research in environmental toxicology since the late 1940s and was among the first ecologists to recognize restoration as a serious challenge, a responsibility, and an important opportunity for ecology, recalls that this entailed crossing a conceptual and cultural watershed and that this took time. The history of restoration, he says, is a bit like the history of the steam engine: Practice came first, then the theory behind it. He began thinking about restoration early on when, as curator of limnology at the Academy of Natural Sciences in Philadelphia in the 1950s, he realized that just getting toxic materials out of an ecosystem did not necessarily result in recovery of the preexisting biotic associations, that something more — active restoration — was often needed. He found himself devoting a lot of attention to this, but also found that it made him something of an odd man out among his academic colleagues. "Ecologists at the time focused on natural systems," he notes. "They ignored systems that were obviously influenced by humans, and pretty much ignored disturbed systems, unless they saw the disturbance as natural — resulting from drought, for example or a hurricane. So the few of us who were working on human disturbances such as pollution were really looked down on. People would ask us, 'How are things down at the factory,' and make comments like that, and if I hadn't been working under the protection of Ruth Patrick, who was chair of the department, it was pretty clear it would have taken me a long time to get tenure."[15]

This would not change, and restoration would not become a serious discipline, until practitioners and ecologists took to learning from, arguing with, mentoring, competing with, and befriending each other, while standing as a body to make the case for what they were doing to a circle of skeptical—or indifferent—outsiders.

This entailed two realizations, achieved and acted on by many practitioners independently. The first was the realization that the old ecosystems *could* be reassembled, at least in some cases, from scratch and on a small scale. The second was that this practice is paradigmatic—not just a game to be played in the academic sandbox or as a display in a public garden, but a mandate—the extreme example of what you have to do to *any* ecosystem in order to keep it on its historic trajectory. This connection was made from both directions, as restorationists scaled up their projects and as managers realized that their attempts to maintain "natural areas" by controlling exotics, reintroducing extirpated natives, and reinstating processes such as burns and hydrological cycles actually *were* restoration and that there were good reasons for framing it in this way rather than in noncommittal terms such as *management* or *stewardship*. As this happened, it also constituted a radical shift in conservation thinking, practice, and expectations. For the better part of a century, natural area conservation had been a strictly defensive operation, the highest expectation for which had been to slow what most supposed was an inevitable decline toward zero. This is the dispiriting sentiment Leopold had encapsulated in his comment that "wilderness is a resource which can shrink but not grow."[16] Of course, as Paul Shepard realized in contemplating the strip-mined landscape at Green Oaks, wilderness *does* grow, simply by definition: If you leave a place alone, that's wilderness. What will usually *not* recover or expand is a particular ecosystem, and a deep sense of this hung over the environmentalism of the quarter of a century that followed the publication of *Silent Spring*. The result was a deep-seated pessimism about prospects for the future of the ecosystems environmentalists regarded not only as models of ecological health but also as emblems and repositories of nature at its purest and most natural.

The prospect—or promise—of ecocentric restoration challenged this pessimism. To the extent it proved feasible in a given situation, it meant that the old ecosystems could gain as well as lose ground. And this not only gave them a plausible future on a planet increasingly dominated by humans, it also changed the sign of human relations with them from negative to positive. That meant, contrary to much of the environmental rhetoric of the period, that perhaps humans, including modern, industrially advanced, even Western humans, actually do belong on the planet. This amounted to a revolution, holding out the possibility of a new kind of environmentalism, one that offered not only a plausible future for the old ecosystems but also the basis for a new kind of relationship with them.

If this meant challenging some of the foundational assumptions of a preservation-oriented environmentalism, it is not surprising that it came not from the centers of environmental thinking and conservation practice but from its margins. It took its first steps not in the national parks or wilderness areas but in small, scattered preserves and on ecologically degraded, even ruderal land, often in urban and suburban settings. At the same time, it was in some ways at odds with the habits and fashions of the academic ecology of the period, notably its preference for "undisturbed" areas as objects of study, its tendency toward reductionism, and its predilection for analysis and modeling rather than synthesis and practice. Its pioneers were on-the-ground types, in many cases amateurs with only a marginal relationship with the academy or with institutionalized land management. Although plenty of professionals were practicing various forms of meliorative land management, such as sustained-yield forestry, game management, and soil and water conservation, by the 1970s few if any were getting paid to do ecocentric restoration. Yet by that time it was, if not yet a clearly articulated idea, at least a practice whose time had come.

Looking around the United States as best we could, mainly by talking with restorationists in various parts of the country, we found projects appearing with increasing frequency in the 1960s and 1970s. These were not the result of any kind of movement, coordinated effort, or even clearly defined idea. Each project was in large part the invention of a single person or small group responding to a particular situation but also, if only half-consciously, to conditions, concerns, and considerations taking shape in the environmental culture of the time. Some were scientists, working in academic settings or for organizations such as mining companies. In any event, what was going on was very bottom-up in an intellectual as well as a sociological sense, not so much the application or reduction to practice of an idea as an action undertaken in response to a situation perceived first as a problem and then as an opportunity, and only later identified and experienced as the enactment of an idea, which came to be labeled "restoration."

In Love with Prairie

Up to now, the efforts at what we are calling ecocentric restoration that we have profiled were historically, professionally, and socially isolated. Although the few that survived more than a decade or so eventually served as models for other projects, it was some time before such projects began to influence each other to the extent that they began to generate a culture of sorts as those involved became aware that they were doing something that was not only new but important and started talking to each other.

In the prairie region this development may be dated to 1960, when Bob Betz, a biochemist at Northeastern Illinois University on the north side of Chicago, joined a field trip to visit a remnant prairie in the southern part of the state and, by his own account, fell in love with the prairie. The reasons Betz gave for this, when he talked with reporter John Berger years later, exemplify the distinction we are making between ecocentric restoration and meliorative land management. What struck him about the prairie was the realization that the bits of prairie he was seeing in out-of-the-way corners, all but lost among the corn and soybeans of rural Illinois, were remnants of something both vast and surprising—not at all like the meadows and old fields people called "prairies" in suburban Chicago, a realization that came to Betz as a "startling revelation." "You got this feeling," he told Berger, "of something that went back all the way for thousands of years. This was what the real vegetation of Illinois was like, not the thing I had assumed."[17] It was not familiar but strange and, in a temporal sense, exotic, suggesting to Betz the possibility of importing it from the past back into the present. In keeping with this, in his account of Betz's early work on restoration of prairies Berger says nothing about their practical value for reconstituting soils or providing habitat for game but simply celebrates this pioneering work as an activity that, like chess or handball, is worthwhile for its own sake—not work, that is, but play, like Leopold's weekend tinkering in Sauk County.

Realizing that bits of the old prairie still existed along railroad rights-of-way, in old cemeteries, and in preserves around and even in Chicago, Betz began seeking them out. He also began taking out exotics, clearing invading brush, reintroducing fire, and introducing missing native species at a few sites. These last he did surreptitiously, sensing that although others might countenance housecleaning efforts such as removal of exotics, they might think that reintroducing species compromised the integrity and naturalness of the remnants.

Betz's work with the cemetery prairies developed into a kind of trademark, and it remains arguably a definitive model for ecocentric restoration. After all, Betz not only aimed at the recovery of a whole historic ecosystem but was actually restoring ecosystems that still existed. In other words, there was something—an altered (or degraded) ecosystem—to restore, and this distinguished these projects from from-the-dirt-up projects such as the one at Madison, which might more accurately be called re-creations.

The same year Betz discovered the prairie, Ray Schulenberg, a horti-culturist at the Morton Arboretum in the western suburb of Lisle, was asked by his superior to create a prairie at the arboretum. Schulenberg, whom Betz had met in the course of his prairie peregrinations and who was propagator and curator of native plants for the arboretum, had tin-kered with prairie years earlier and took on the task enthusiastically. Meticulous to the point of obsessive in his work, Schulenberg started as Henry Greene had done at Madison a decade and a half earlier, using a labor-intensive, horticultural approach to prairie restoration that involved hand-planting of banded plants and weeding by interns working on hands and knees with linoleum knives. The result within a few years was several acres of spectacularly high-quality prairie vegetation. Like Greene, Schu-lenberg gradually shifted to less labor-intensive methods based on broad-cast seeding as he scaled up the project, which now covers roughly 40 hectares.[18]

A decade later Betz collaborated with The Nature Conservancy's Bob Jenkins to undertake re-creation of prairie on 260 hectares of former farm-land inside the proton accelerator ring at Fermi National Laboratory in nearby Batavia. Eager to scale up, Betz developed seed mixes based on the idea that the best way to restore prairie was to mimic succession by intro-ducing species in waves of increasingly "conservative" species. He also used mechanical harvesters to gather thousands of kilograms of seed and spreaders mounted on all-terrain vehicles to distribute it, developing an agronomic approach to the planting effort that complemented Schulen-berg's meticulous horticultural approach. Betz made his first plantings in 1974 and added parcels on a scale of tens of hectares in subsequent years, making the circle of land inside the accelerator ring a patchwork of prairie in various stages of development that has since provided opportunities for research unmatched anywhere.[19] By 1992, the project had expanded far outside the ring to encompass nearly 400 hectares, and Bill Jordan recalls being a bit chagrined when, probably some time in the early 1980s, he heard about the project and realized that although the UW–Madison Ar-boretum's prairies might still be the oldest, he could no longer claim that they were the largest restored ecological communities in the world.

Together, these projects, first at Green Oaks and then a few years later at the Morton Arboretum and Fermilab, marked the beginning of restora-tion as a coherent movement in the Chicago region. Not only did all three have dynamic, charismatic leaders working in secure, supportive institu-tional settings, but they were highly visible, drew public attention, and, just as important, began to build up a cadre of committed volunteers who provided the foundation for the grassroots-based restoration culture that would play a crucial role in the discovery of the value of this work. Moreover, those involved were talking with one another, developing an awareness that what they were doing was important, and beginning to define themselves as restorationists. They were also promoting the idea of restoration. As a result, a small but increasingly coherent restoration culture began to form, and projects began to proliferate. Bob Betz's work, for example, took on the character of a crusade as he ransacked Illinois for prairie remnants and cajoled town councils and cemetery committees into endorsing his efforts to restore and maintain them. Within a few years of starting the project at the Morton Arboretum, Ray Schulenberg was working with David Blenz to restore prairie on a 2-hectare site at nearby Camp Sagawau, a unit of the sprawling Forest Preserve District of Cook County, which emerged over the next few decades as a proving ground and influential showcase for community-based restoration efforts.

Overall, the initiatives of this handful of practitioners had brought restoration to a new threshold of realization, giving it a fertility—or infectious character—it had not had before, even when represented by high-visibility projects such as the one at the UW–Madison Arboretum, where the scientists involved had little interest in promoting restoration as a cause and, far from promoting community participation, actively discouraged public use of the property.

Chapter 7

Realization II: Taking Hold

Although the tallgrass prairies of the Midwest would retain the reputation as exemplars of ecocentric restoration they had acquired early on, projects were taking shape on behalf of other kinds of associations in other parts of the country during this period. Indeed, what we find as we look for relevant initiatives in different parts of the United States beginning in the 1960s is a kind of convergent evolution, with practitioners working in a wide range of systems and under a wide variety of ecological, institutional, professional, social, and economic conditions, moving more or less independently toward the idea of ecocentric restoration.

This being the case, following the development of restoration from this point through the past few decades puts us in a position to ask some interesting questions about the idea of ecocentric restoration and how it emerged—or didn't emerge—under various cultural and ecological conditions. If, for example, ecocentric restoration entails a measure of impracticality, including costs that will not be recovered, it necessarily has about it something of the amateur spirit. And if for this reason it was invented by amateurs such as Bob Betz or by professionals working on the margins of their disciplines or institutions, then what happens to it when it "goes commercial" or is undertaken as an obligation imposed from the outside by regulations? Similarly, what happens to it—or, more accurately, what version or idea of restoration emerges—when the objective of returning an ecosystem or landscape to a previous condition conflicts with economic aesthetic or other human interests?

What we find as we look back at the development of restoration in different parts of the country over the past two or three decades is that the efforts undertaken under the label *restoration* typically reflected all these factors. Many were essentially resource conservation efforts undertaken for frankly utilitarian ends. At the same time, we find growing numbers of projects that clearly were inspired by, or at least reflect, the notion that there may be some value in restoring "all the parts" of an ecosystem, even if that larger and more generous objective has to be smuggled in, disguised in the conventional language of utilitarian self-interest. Indeed, this tension between unstated bio-altruistic aims and the claims of utilitarian self-interest has been a consistent subtext of the more ecologically ambitious forms of restoration as restoration has moved beyond the academic sandbox and into the "real world" of regulations, competition, financial constraints, and concerns about customer satisfaction.

In the Northeast, for example, the idea of ecocentric restoration had less salience than it had in the Midwest because trees, at least, if not the old forests, reestablish themselves on abandoned land in a way that prairies do not.[1] Beyond that, when the idea was applied to ecosystems other than the politically disenfranchised prairies, it had to make its way in the context of protective or prescriptive legislation, the effects of which were complicated. Legislation such as the Clean Water Act of 1972 created a market for restoration of clean water. However, that can be achieved without reference to any kind of historic ecosystem. Besides that, legislation creates a situation in which objectives are compliance rather than performance, so that those involved may be doing—or paying for—restoration efforts not because they want to but because they are required to. Under these conditions, ecocentric restoration is likely to lose ground to more "practical" considerations that may or may not be consistent with its aims.

"When that happens," says restorationist Leslie Sauer, "the question is usually 'What is the least we can get away with?' A set of standards intended as a floor beneath which you can't go tends to become a ceiling no one wants to exceed." Sauer, who was a founder of Andropogon Associates in Philadelphia and one of the first generation of practitioners who were feeling their way toward ecocentric restoration as a profession in the 1970s and 1980s, recalls how the regulations coming online during this period both created opportunities for the organization and tended to subvert its mission. Sauer herself had worked with landscape architect Ian McHarg, and Andropogon (named for a genus of grasses abundant in North American grasslands) was set up to explore ways of implementing his idea of "designing with nature." Ecocentric restoration is the extreme version of this idea, and Andropogon was promoting it even though in the organization's early years the word itself was not current, and Sauer and her colleagues described what they were doing simply as "ecological planning and design."

"When we started out, there was no such thing as getting a job restoring a native landscape," Sauer recalls. Early on, putting one toe into the regulatory environment that was taking shape in the 1970s, Andropogon undertook restoration of a small spatterdock wetland that had been illegally filled—the first restoration mandated by the Philadelphia District of the Corps of Engineers. Sauer and her colleagues were frustrated, however, by a lack of agency oversight and standards that they thought were not related to environmental quality.

As for regulations in general, Sauer and her colleagues found that although regulatory requirements often brought clients to the firm, they also made it harder to make the case for the higher-end sort of work Andropogon wanted to do. Sauer recalls being aware of some excellent projects in landscaping and also in areas such as mine reclamation early on, but she says that efforts such as these were often "subverted." When regulations came online, enforcing standards that fell short of objectives that practitioners had defined for themselves, they often undercut both initiative and arguments for these more ambitious aims.

"That just added to the pressure we had to work against," she recalls. So Andropogon basically opted out of the mitigation process for decades and looked for other ways to practice what amounted to ecocentric restoration. Occasionally they did this on a pro bono basis or as a class project. For example, Sauer recalls working with a class at the University of Pennsylvania to restore native vegetation in a woodland glade in Philadelphia's Fairmount Park in memory of McHarg's recently deceased wife, Pauline.

Even in their for-profit projects, however, the Andropogon team did its best to build regard for the whole ecosystem into every project they took on. What they discovered, and learned to take advantage of, was the distinctive appeal of the idea of actually restoring historic ecosystems or landscapes.

"We learned that people really like the idea of restoration," Sauer says. "It is a far bigger vision than mitigation, and it has real appeal, even to the corporate world, which understands the PR value it has and buys into it as long as it is not more expensive. So we tried to capitalize on that. The fact is you can't set standards high enough," she says. "I mean, you can't start with impossibly high objectives. But once people start out with more modest aims, they often get carried away and wind up doing amazing things, and they often get far more demanding and meticulous than any regulations or even company guidelines could possibly be." This, she says, generally has less to do with the site or client than with "a few people who really see the paradigm shift." She notes that even in its "less inspired forms, the attempt to restore a landscape" is enlightening and seems to change people. "At one point, back in the '80s," she recalls, "we had a contract with New York City to do some work on the Fresh Kills landfill on Staten Island. The aims were entirely utilitarian—to cap a landfill—but they weren't using enough soil to sustain a landscape through a drought. So we moved the project toward a more holistic, ecocentric approach, which of course produced a much better result in every way. So when the man who was managing that project moved on to his next project he had a much more sophisticated understanding of what was going on. And much higher standards. At one point, for example, the soil the City was sending down for fill was larded up with nutrients, which were inappropriate for the project, and during one period these guys turned away every truck that came down there for two entire years. Can you imagine that happening on a commercial job being carried out under conventional regulations? Well, they were just determined to re-create an eastern grassland out there, and they were doing what they had to do. And they did that on their own; this was entirely an internally-defined goal."[2]

Wetlands

Clearly, regulation entails tradeoffs that typically favor meliorative over ecocentric restoration. But, as Andropogon's experience makes clear, it also offers opportunities, and practitioners who are committed to pushing their work in the direction of this form of restoration have learned to take advantage of these, finding ways to piggyback ecocentric projects on the backs of resource-oriented projects in a kind of ecological black market, or what the National Oceanic and Atmospheric Administration's Peter Leigh has called "closet ecocentrism."

Wetlands are a good example. For one thing, they not only have short-term economic value as habitat for fish and waterfowl, they are also subject to a bodyguard of regulations, notably the Clean Water Act and the Swampbuster provisions of the 1985 Farm Security Act, *and*, again in contrast to the prairies, they do not generally lend themselves to tinkering by amateurs. For all these reasons, the early wetland restoration efforts, undertaken around the turn of the twentieth century, were motivated by concerns about natural resources—usually fish or waterfowl—and were carried out mainly by professionals, often with the support of a resource-oriented nongovernment organization such as the Isaac Walton League or Ducks Unlimited, in partnership with a state or federal conservation agency.

This pattern held true half a century later, when managers began to take an interest in the possibility of restoring wetlands for their own sake. Here again, as in the case of forests, this development was preceded by a long history of management motivated in large part by utilitarian consid- erations.[3] In fact, plantings of mangroves and salt marsh species, in some cases covering hundreds of hectares, were undertaken as early as the late 1800s in Europe, China, Australia, New Zealand, and the United States to slow erosion, reduce channel siltation, and reclaim land for agriculture.[4] On the East Coast, Frederick Law Olmsted picked up on this idea and moved it a step closer to ecocentric restoration by integrating hydrological considerations with aesthetic ones in a project he designed for Boston's Back Bay in the 1880s.[5] Despite this early initiative, however, those cur- rently involved in wetland restoration point to much later projects, moti- vated by explicitly utilitarian considerations, as the immediate context for the development of a more ecocentric approach to the restoration of tidal wetlands.

People we talked with trace the beginnings of coastal restoration on the East Coast to the work of William W. Woodhouse Jr., an agronomist and soil scientist at North Carolina State University who in the 1960s experi- mented with ways to establish vegetation on coastal dunes subject to storm damage and on dredged materials from Army Corps operations. Although much of this work was essentially land reclamation involving attempts to stabilize dredged materials with *Spartina alternifolia*, it did include an ecocentric dimension. Steve Broome, who did graduate work with Wood- house in the early 1970s and is now a professor of soil science at NC State, notes that at first the motives for this work were strictly utilitarian but that this began to change in the 1980s as "restoration" and "mitigation" gained currency and researchers began to pay more attention to the value of this kind of work as a way not only of reclaiming severely disturbed sites but of actually creating and restoring tidal wetlands. Soon, he notes, the Army Corps was talking about restoration of habitat, was taking an interest in us- ing dredged material to create and expand tidal marshes, and was explor- ing soft alternatives to concrete as a way of stabilizing shorelines, a devel- opment that led directly to the Corps's ambitious restoration work on Florida's Kissimmee River, which began around the same time.

Christopher Craft, now at Indiana University, who did graduate work with Woodhouse in the mid-1980s, adds that the mitigation tradeoff, often deplored by environmentalists, actually helped here because the idea that restoration might be used to compensate for damage to existing wetlands required, at least in principle, that the restored wetlands be the ecological equivalent of those they were intended to replace. Even more than the Clean Water Act a decade earlier, this provided a regulatory mandate for high-quality restoration.

In the meantime, a similar story was taking shape along Florida's Gulf Coast, where Roy R. "Robin" Lewis III, then a young biology professor at Hillsborough Community College in Tampa, found a utilitarian "excuse" for undertaking path breaking work on restoration of what at the time was regarded as a useless ecosystem—coastal mangrove swamp—around Tampa Bay. Mangroves had declined around the bay over the years because of a combination of declining water quality and development of high-priced shoreline property. Because they were generally regarded as mosquito-breeding nuisances no one much cared, however, and when Lewis proposed a restoration effort on their behalf he got little support, even from fellow biologists. "Everyone said the mangroves were worthless, even in biological terms," he recalls.

Lewis and his students did the project anyway. Censusing fish populations in the mangroves, they found that they provided habitat for small fish that were an important food for shorebirds and wading seabirds such as heron. They were, in other words, of great interest not only to ecologists but to birders, a connection between self-interested and other-regarding considerations that turned out to be useful in promoting the aims of ecocentric restoration. Lewis began a series of experimental mangrove restoration projects around the bay in 1974 and soon realized that mangrove restoration was a simple matter if properly done. But if the biologists and developers had been skeptical about the idea of restoring mangroves, Lewis now encountered skepticism about the whole notion of restoration. When Lewis, aware of the wetland restoration efforts getting under way elsewhere, proposed a symposium on the subject, the college was wary. "They were afraid of the idea," Lewis recalls. "They said we could use a room, but they wouldn't let us list them as a sponsor. So we went to Florida Audubon with the mangrove–fish–shorebird connection we had documented. The people there liked that, and they wound up sponsoring the conference."

The conference, ambitiously billed as "The First Annual Conference on Restoration of Coastal Vegetation in Florida," held at the college in May 1974, was the first conference anywhere devoted specifically to coastal wetland restoration and the first in a series of conferences that have been held annually right down to the present. Lewis himself found the work of restoration so rewarding that he left academia in 1980 to found a firm, now called Lewis Environmental Services, that offers training and does design and wetland restoration work internationally.

Rattlesnakes and Poison Ivy

Andy Clewell, who left a job teaching botany at Florida State University in 1979 to do restoration work, tells a similar story of leaving academia to do restoration and of working creatively to find ways to leverage holistic projects out of legally mandated reclamation efforts, in his case reclamation of lands devastated by surface mining for phosphate near Lakeland, Florida. There is a lot of restoration work going on in northern and central Florida, he says. But, constrained by strictly worded, inflexible permits that typically favor speed, simplicity, and shortcuts rather than patience and the attentive adaptive management needed for high-quality work, very little of it qualified as ecocentric restoration. His first chance to do what he calls Madison/Chicago-style restoration in a serious way came in 1997, eighteen years after he started his business, when he directed restoration of wet prairie on 753 hectares of abandoned pine plantation at Old Fort Bayou near Ocean Springs, Mississippi. That project provides a good illustration of the complex relationship between regulations and restoration efforts with objectives that exceed mandated standards. On one hand, it was carried out as a "deposit" in a mitigation bank set up by the Mississippi Nature Conservancy to provide developers with a way to compensate for environmental damage, and so it was ultimately dependent on the market for restoration created by the regulations. On the other, it was carried out by a nonprofit organization, which set goals in keeping with its mission of preserving "all the parts." Besides this, Clewell notes that the lead regulatory agency itself, the Mobile, Alabama District of the Army Corps of Engineers, had not yet developed standards for this kind of project and left the job pretty much up to Clewell and his colleagues.[6]

Given this chance, they went for all the parts, good, bad, and ugly. So when Clewell noticed poison ivy, a native to the ecosystem, on the site he left it alone, and when a diamondback rattlesnake showed up he regarded it as a transspecies endorsement of the project.

Of course rattlesnakes are one thing, and government bureaucracies are another. Clewell, who has worked on regulated projects for three decades, suggests that the success of the Old Fort Bayou project was due in part to the trust the Army Corps extended to the Nature Conservancy and the opportunity the conservancy in turn gave him to undertake the kind of tinkering he thinks is necessary for high-quality restoration. "It really works best when the agency doesn't know what it's doing, and depends on the contractor to set the standards for a project," he says. "But then if the project more or less works out, the agency makes it a basis for a set of cookbook rules, which they impose on other projects, basically taking the initiative away from the practitioner and making it harder for him to respond to the distinctive conditions on a site."

A similar mix of circumstances and motives, though with its own distinctive flavor, has prevailed along the West Coast. There the development of restoration has reflected the influence of a grassroots culture drawing inspiration from the counterculture as members of Charles Reich's "greening of America" generation moved back to the land in the aftermath of the war in Vietnam. As in the East, projects mandated by regulations, though more often than not falling short of the aims of ecocentric restoration, have nevertheless often included elements that had previously been ignored. This was especially true in the case of wetlands, which, like the midwestern prairies and unlike more charismatic ecosystems, such as rivers, creeks, and redwood forests, have small constituencies. John Stanley, a restoration ecologist with WWW Restoration in Boulder Creek, California, points to early work of ecologist H. Thomas Harvey, with whom he worked at San Jose State College (now University) in the late 1960s and early 1970s. Harvey undertook what Stanley believes were some of the earliest restoration projects in the region, including work with giant sequoia forests in the 1960s and restoration of the 40-hectare Faber Tract Marsh in Palo Alto on San Francisco Bay, which Harvey began with his San Jose State colleague Howard Shellhammer in 1971.[7] Stanley notes that Harvey was interested both in the prospect of reversing environmental degradation and in taking advantage of the opportunity for some ambitious research projects, but both of these concerns reflected concern for and interest in the ecosystem for its own sake.

Projects such as these did provide ecologists with opportunities to undertake research that would be difficult or impossible to carry out in any other circumstances. Ecologist Joy Zedler, for example, undertook pathbreaking research on the restoration of tidal wetlands in and around San Diego Bay in the 1970s. She took advantage of large-scale restoration efforts mandated by mitigation agreements under the Endangered Species Act to initiate long-term research that took into account a wide range of habitat parameters that are difficult for amateurs to document, including functions such as nutrient cycling. This attempt to define the critical parameters that define habitat for target species resulted in important insights into ecosystem structure and function—early examples of what would later be called restoration ecology. It also provided ammunition for arguments for additional restoration work, including purchase of Famosa Slough, adjacent to San Diego's Mission Bay, in the 1980s, followed by what Zedler describes as a "magnificent" project in which volunteers got the city to control urban runoff, then undertook to remove trash, control exotic vegetation, and replant native species. That and other citizen-based efforts to restore coastal wetlands provided critical momentum for a lawsuit that led to the abandonment in mid-construction of an interchange on Interstate Highway 5 in Chula Vista, just south of San Diego.

Overall, it is clear that top-down government influences in the form of legislation such as the Clean Water, Swampbuster, National Environmental Protection, and Endangered Species acts have played an important, if at times ambiguous role in the development of restoration in its various forms. This is especially evident in the case of wetlands, for the simple reason that, in contrast with prairies, wetlands don't lend themselves to isolated projects, à la Henry Greene toiling with a pail and shovel on his site in Madison. Arnold van der Valk, an ecologist at Iowa State University who has been doing research related to the ecology and restoration of wetlands in the prairie pothole region of the northern Great Plains since the 1970s, points out that wetlands are connected with their environment by the water that defines them, so that anything you do to a wetland affects your neighbor and entails political considerations, which become increasingly fraught as you work your way down the watershed. As a result, he points out, the drainage of vast areas of the Midwest during the period of European settlement depended on the development of a political and regulatory infrastructure capable of funding and coordinating these efforts, and this has also been true of attempts to undo this work on selected sites in order to restore them to their presettlement condition.[8]

Van der Valk mentions the Swampbuster bill, which denies subsidies to farmers who disrupt wetlands, noting that it represents the first time that subsidies were made contingent on any form of environmental stewardship. This led to programs that allowed farmers to put low-lying land into conservation reserve programs, some of which promoted the restoration of pothole wetlands, an important step in the direction of restoration on private land in midwestern agricultural areas. He also notes, however, that work carried out under such programs has usually been driven almost exclusively by utilitarian considerations, with water quality succeeding waterfowl habitat as a priority in recent years, and that it is usually not well coordinated, limiting its ecological value. In one study, for example, his group found that the value of wetland restoration efforts in the "Iowa Great

Lakes" area in the northwestern part of the state was limited because only about 15 percent of the water delivered to the area as precipitation goes through a restored wetland on its way into the lakes.

As for ecological quality at the community level, progress in that direction has often been limited by the utilitarian concerns driving the relevant legislation, perhaps aided and abetted by a lingering Clementsianism. Van der Valk says that for years the prevailing view was that if you restored the hydrology, the community would come back on its own. This notion obviously saved a lot of time and trouble, and Van der Valk notes that, somewhat ironically, it was encouraged by misapplication of his own work on the recovery of wetlands from seedbanks. That work showed that vegetation lost as a result of high water over two or more years recovered well from the seedbank but also that seedbanks on *drained* lands were depleted of most species after about twenty years — an aspect of this work that the authorities largely ignored. The result, of course, was that areas that were reflooded decades after being drained generally developed stands of cattails and reed-canary grass — suitable for waterfowl habitat, but a long way from the historic vegetation of these sites.[9]

Similarly, on the West Coast, John Stanley notes that by far the majority of projects — "maybe 95 percent" — were aimed at meeting mitigation requirements that fell well short of the aims of ecocentric restoration. On the other hand, legislation such as the Endangered Species Act, with its more stringent requirements for restoration of ecosystems tailored to provide habitat for one or more rare species, has pressed restorationists explicitly in this direction, beyond market considerations and into the impractical business of acting on behalf of "useless" species and the sometimes obsolete ecosystems that provide their habitat. A prime example was early work carried out under the auspices of The Nature Conservancy in partnership with Ducks Unlimited to restore riparian areas along the Cosumnes River in the Sacramento Valley in the late 1980s. The aim was to provide habitat for the yellow-billed cuckoo and other birds associated with broad expanses of floodplain riparian habitat. In this case, ecologists had identified a corridor of riparian forest at least 200 feet wide as a critical factor defining habitat for the cuckoo. Achieving this could be costly, however, and although a client aiming to meet regulatory requirements might argue for a narrower corridor, the conservancy accepted this criterion and developed a number of low-tech means of achieving it on a limited budget.[10] Here the conservancy's commitment to "preservation" of the old association, given teeth by the Endangered Species Act, turned what could have been mere compliance into a real ecocentric restoration project.

Driven in part by legislative requirements, wetland restoration has become an important component of conservation practice over the past few decades and also a major business. At the same time, questions about the ecological quality and historic authenticity of restored wetlands have gained more attention.[11]

The Bay Area

Old-timers now in their fifties and sixties recall similar developments in the San Francisco Bay Area during this period—developments that matured rapidly to make the region an epicenter of restoration activity. Practitioners such as John Thielen Steere, who takes a long view, attribute the development of restoration in the Bay Area to the spectacular landscape itself; the rich biodiversity resulting from a high-relief topography; its mild, Mediterranean climate and proximity to both marine and aquatic systems; and a tradition of nature mysticism, dating back at least as far as John Muir, that found expression in later generations of Beats, Hippies, and other questioners of human hegemony over nature.

Robin Freeman, who has been involved in environmental activities in the area since the early 1970s and now heads the Environmental Management and Technology Program at Merritt College in Oakland, traces modern, scientific interest in the health of the bay ecosystem to the late 1950s, when theatrical producer John Reber proposed a scheme to fill in large areas of the bay, replacing the whole estuary with two freshwater lakes and a system of dams and channels in order to create land for development and expansion of military bases, reservoirs, roads, and airports. The Reber plan was abandoned in the 1960s, but the prospect of such large-scale meddling with the defining feature of one of the world's most distinctive and widely recognized landscapes raised public awareness that the future of even such an act of God as San Francisco Bay is not something to take for granted.[12]

This concern focused on preservation, a first step toward interest in restoration, which began, those we talked with agree, with two developments. The first was in the bay itself and its estuaries, when scientists began taking an interest in their ecology and management. The second was when citizens, defending creeks and streams running into the bay from entombment in concrete-lined channels—the conventional way of dealing with urban waterways at the time—realized that they could not only save them from burial but could disinter them, and turned to the scientists for advice on how best to do so.

Here again, repeating the experience of George Wright and his colleagues in the national parks decades earlier, the objectifying perspective of science played a crucial role. As Robin Freeman points out, like Florida's mangroves and the tallgrass prairies of the Midwest, the estuaries, tidal marshes, and willow groves that fringe the bay were of little interest to most citizens, who regarded them as waste areas and mosquito breeding grounds best filled and converted to other uses. Scientists, whose interests included the biota of the bay as well as "practical" matters such as erosion and flood control, naturally had a different perspective, which eventually informed attempts not only to save but also to resurrect and restore remnants of historic ecosystems, both in the bay itself and in the uplands surrounding it. This began in the bay in the 1950s and early 1960s, when the U.S. Geological Survey, partly in response to questions raised by the Reber plan, undertook studies of the floor of the bay to test ideas about plate tectonics. In the process, Freeman says, the scientists involved realized they knew very little about the bay, and they undertook studies of its geomorphology, hydrology, and ecology that led to the publication of the *San Francisco Estuary Project: Comprehensive Conservation and Management Plan*, the first edition of which came out in 1993.[13]

The *Comprehensive Plan*, which provided guidelines for restoring ecosystem health and limiting the effects of development on the bay, has served, Freeman says, as the "policy bible" for management of the bay, providing a foundation for restoration efforts higher in the watershed. That story began when bulldozers began arriving in some of the better-heeled neighborhoods around the bay, on a mission to channelize waterways. In response, in 1961 Sylvia McLaughlin, who lived in Berkeley, spearheaded the Save San Francisco Bay project to consolidate local resistance to this form of land management. She later played a key role in the formation of Citizens for an East Shore Park, which arranged the purchase of land around the bay, turning it over to conservation organizations and agencies to act as stewards. These initiatives in turn contributed to the formation of the Bay Conservation and Development Commission in 1965. This organization, which had legal and governmental authority to approve or deny applications for development, became the first coastal protection agency in the United States when it was made permanent in 1969.

All this is prologue to the emergence of restoration in the area. Those we talked with consistently identified two waterway restoration projects as seminal events in this development. The first of these was a project on Wildcat Creek in North Richmond, spearheaded by Ann Riley, who studied hydrology with Luna Leopold, Aldo's son, at UC–Berkeley in the 1970s and 1980s and has since played a leading role in waterway restoration efforts nationwide. In 1982, Riley joined community leaders in objecting to a plan by the Army Corps of Engineers to channelize roughly 3 kilometers of Wildcat and San Pablo creeks, which they thought was inconsistent with the "model cities" plan North Richmond had developed in the 1970s. At the same time, a coalition of African American community leaders and the fledgling Urban Creeks Council joined other environmental and community organizations to acquire funding and worked with hydrologist Phillip Williams to develop a new concept in urban flood control, based on natural channel and floodplain geomorphology. What ensued was a David-and-Goliath-style controversy, but the "people's plan," which was given a boost by the Endangered Species Act, prevailed, and its implementation began in 1986.

Around the same time, in 1979, Carole Schemmerling, a resident of Berkeley, heard a talk on creek restoration efforts in the counties north of the bay. These were not urban projects, but the speaker, historical geographer Gray Brechi, noted that similar projects could be undertaken in urban areas. Schemmerling, who had not been aware of the prospect of stream restoration before, found these accounts of the resurrection of waterways immensely exciting. Shortly after this, she found herself chairing the Berkeley Parks Commission, which was considering how to allocate funds from a bond initiative to provide funding to develop parks in neighborhoods that had none, and she spent some time scouting the city for likely projects.

One proposed site was two blocks of abandoned railroad right-of-way that included a stretch of Strawberry Creek that had been culverted since 1912. The city was planning to create a park at the site but not to daylight the creek. So Schemmerling organized a public hearing and recruited David Brower, then at Friends of the Earth and widely recognized as a leading advocate for the environment, to attend and to support a plan to restore the creek rather than reburying it. Brower did so, and the plan was approved. It was carried out by Doug Wolf, a landscape architect who, in contrast with Ann Riley, already at work on Wildcat Creek, knew nothing about hydrology but, Schemmerling says, "got it exactly right." Wolf did so by hunting up clues to the original course of the creek, then restoring its meanders and establishing a floodplain that has protected the area from flooding in every storm event since, including an El Niño event in which the creek overtopped its bank and the re-created floodplain handled the overflow.

Was this ecocentric restoration? Yes, but with the usual admixture of self-interest. The emphasis in both projects was functional, because both

Schemmerling and Riley realized that they had to make the case for them on "practical" grounds, which meant basically two considerations: erosion control and flood control, the last being the leading factor motivating urban creek restoration projects in California, Riley notes. As with the wetlands on the East Coast, the idea that this sort of functional reclamation might also provide a foundation for reassembly of actual communities came later. "Ann said it first," Schemmerling recalls. "It dawned on us that there are other reasons for these projects, that they could include recreation of habitat, too." However, this had little weight with the authorities until the 1990s, when the idea of protecting endangered species was working its way into the bureaucracy, and then, of course, proponents leaned on it heavily in making their case for additional projects. In the meantime, there was pressure in the direction of ecocentric restoration from figures such as Brower, who brought into the conversation the preservationist's sense of the value of the old ecosystem, mixed at times with ambivalence about the prospect of an artificial version of nature. Schemmerling recalls running into Brower on a visit to Yosemite when the Strawberry Creek project was well under way and asking him whether he'd visited it. "No," he said, with what she recalls as "mock severity," "And I'm not going to until the whole creek is open and the salmon are back"—salmon that had not been seen in the creek since 1932 (and still haven't as of 2010, Schemmerling reports).[14]

Not long after Schemmerling and Riley met, Riley mentioned that another project was under way on Glen Echo Creek in Oakland, and so, thinking that three projects represented some sort of critical mass, they started the Urban Creeks Council, which promotes the daylighting of waterways in the region, an activity Schemmerling characterizes as "addictive."

Those we talked with concurred that these two projects were landmarks in the emergence of restoration in the Bay Area—what Christopher Richard, curator of aquatic biology at the Oakland Museum of California, characterizes as "poster children"—valuable despite their ecological defects and limitations, as symbols and models that have inspired the projects that have since proliferated in the Bay Area.

Like the epiphany Schemmerling recalls regarding the possibility of restoring both habitat and ecosystem services, this idea of poster children is an important one as far as the discovery and realization of restoration are concerned. Robin Freeman, for example, notes a succession of projects or events, including the projects at Wildcat and Strawberry creeks, that stand out in retrospect as having importance beyond the purely ecological or aesthetic because they dramatized or articulated some aspect of the idea of restoration. Freeman stresses, however, his impression, echoed by others we talked with, that the real threshold in the cultural discovery of restoration in the Bay Area did not come until the late 1980s, when a series of events led to a widely shared realization of its distinctive value—the social and political equivalent of Ann Riley and Carole Schemmerling's experiences on Wildcat and Strawberry creeks a decade earlier.

Chief among these, in Freeman's view, were a pair of conferences that dealt exclusively with restoration: the "Restore the Earth" conference, organized by ecologist and writer John Berger and held in Berkeley in 1988, and the first annual conference of the newly formed Society for Ecological Restoration, held in Oakland the following year.

Both events, together with the launching of the journal *Restoration & Management Notes* (now *Ecological Restoration*) in 1981, brought people and ideas together in a way that Freeman sees as crucial for the realization of restoration as a conservation strategy. "I think this was the watershed for restoration in this area," Freeman says. "I know I hadn't thought about restoration as a whole before then. [East Park Bay naturalist] Tim Gordon and I had worked with kids planting native plants along Wildcat Creek as early as 1973. But I didn't see that as a distinctive form of land management with important implications for the environment. After those meetings I did, and I think that was true for a lot of people. I know that I and my colleagues at Merritt College, many of whom had been doing relevant work for quite some time, went back and started planning a program to train leaders for the kind of work people had talked about at these conferences. Others launched similar initiatives, and the result was the development of a restoration culture that has really started to affect the environment in the Bay Area."

Cascadia

Further north, the development of restoration in the Pacific Northwest provides an especially interesting study in how the distinction we are making between ecocentric restoration and meliorative land management has worked out in practice. Factors there have included the cultures of pre-Columbian peoples and their efforts on behalf of recovery of the ecosystems they have inhabited and shaped for millennia; a postcontact culture reflecting the experience of early pioneers, who found themselves living in a setting they tended to experience as edenic; and the simple fact of so many living so close to so much nature. Some emphasize the influence of the counterculture that flourished in the region beginning in the 1960s and 1970s when thousands, disaffected by the events of the Vietnam War era, moved into the region seeking alternative ways of living and relating to other people and to nature. This provided the basis for a culture and a sensibility that has been articulated in a rich literature.[15]

Overall, the result was a rich mix of bottom-up, top-down, and end-on efforts that shaped the restoration culture of the region. Although a regional mythology attributes to the area a distinctive sensitivity to nature,[16] some veterans of the environmental struggles of the past few decades are skeptical. "There was just no restoration in the Northwest earlier on," says Billy Frank Jr., a Nisqually, who is chair of the Northwestern Indian Fisheries Commission, headquartered in Olympia, Washington, and has been working with tribal people on conservation efforts in the region for four decades. "It was just management for fish and game, until we began to assert our treaty rights."[17] "I don't think we have been any better than anywhere else," comments Dean Apostol, a landscape architect and coauthor of an encyclopedic book on restoration in the region.[18] "It's just that we got out here later, and had more nature to get through. And we were just mining it—cutting trees, damming rivers, engineering salmon—at least until the 1970s. I think things began to change around that time as environmentalism kicked in and provided a broader context for the work being done by the tribes, by the back-to-the-landers and just plain people who were concerned about what they saw going on with the fisheries and in the clearcuts and so forth."[19]

An early step in that direction was a 1974 decision by the U.S. District Court in the case of *U.S. v. Washington*, which affirmed tribal rights to harvest fish in areas ceded by treaties in 1854 and 1855 and mandated tribal participation in creation of a comprehensive management plan for Puget Sound. This opened the way for the tribes to begin the kind of integrated, ecosystem-scale restoration effort needed to restore habitat for the salmon that are central to their economies and their cultures. It also initiated a succession of decisions supporting tribal projects such as restoration of shellfish habitat and removal or redesign of dams and culverts to reconnect anadromous salmon with their breeding habitat. This, Frank says, has led to some striking successes; he mentions ecosystems that are now "pretty well back together" in the Snohomish and Snoqualmie watersheds in western Washington, fish returning to areas in Puget Sound that had been lethal for decades, and the cleaning up of waters that had been so toxic that anglers had used them to clear their boats of barnacles.

In the meantime, others were fomenting change as well, with the back-to-the-landers and others evolving into a constituency for the environment that increasingly exerted pressure on agencies such as the U.S. Forest Service and the Bureau of Land Management, which control roughly half the land in the region, to move beyond their traditional, resource-oriented method toward practices aimed at ensuring the well-being of whole ecosystems. Veterans of that effort such as Jerry Gorsline and Tom Jay recount a history of bottom-up conservation, as young, idealistic urban expatriates moved into the region, grew disillusioned with conservation-as-usual, and began recruiting allies in state and federal agencies to the cause. Gorsline recalls working as a tree planter on reforestation projects early on, only to realize the work was serving only the interests of industrial forestry, and then turning to efforts to, as he puts it, "reform the forest industry."[20] Similarly, Jay and his wife took on the task of restoring a defunct salmon run near their home in Chimacum on Washington's Olympic Peninsula. Devising means to rear fish and regenerate populations, recruiting colleagues, and working with field biologists from the Washington Department of Fish and Wildlife to gain the support of a skeptical agency, they wound up creating a watery counterpart to community-based restoration programs such as those taking shape in places such as the Midwest and the Bay Area during the same period.

This kind of intense, hands-on activism, repeated in projects all over the region, resonated with a new generation of conservation professionals coming out of schools of forestry and game management imbued with the spirit of the environmentalism that was taking shape at the time. As a result, the culture of the agencies began to change. Both Apostol and Gorsline see a meeting convened by the Clinton administration and held in Portland in 1993, as a watershed in the emergence of restoration in the region. The resulting *Forest Ecosystem Management Assessment*[21] was the first regional-scale conservation plan in the country. It represented at least a partial resolution of conflicts over the use and management of resources typified by the timber-versus-spotted owl battle and has encouraged the development of restoration in the region.

The vigor of the restoration culture that has taken shape in the region is evident from the overview provided by *Restoring the Pacific Northwest*, which Dean Apostol wrote with Marcia Sinclair. The book, which provides an in-depth overview of restoration other regions might envy, details restoration efforts at every scale and on behalf of every kind of ecosystem, from tidal marshes to high-altitude systems, reflecting every conceivable mix of motives. These have substantial support from a healthy mix of private land owners and federal, state, and local government agencies. Apostol and Sinclair note that all four federal land management agencies active in the region—the U.S. Forest Service, Bureau of Land Management, Fish and Wildlife Service, and National Park Service—have embraced restoration as an integral part of their mission. They also note that the practice of restoration, increasingly integrated into K–12 curricula and community-level land management practices, has played a key role in leveraging support for a range of ambitious—and in some cases costly—restoration efforts.

Apostol sees this rich array of restoration projects as a valuable case study in the historical relationship between ecocentric and meliorative restoration. Much of this work has been carried out on behalf of ecosystems representing the happy situation in which meliorative and ecocentric restoration largely coincide. For example, salmon are an indicator species, the comings and goings of which both depend on and influence almost every component of the ecology of the region. As a result, to have salmon you have to have pretty much the whole ocean–estuary–riverine–forest ecosystem they inhabit. Here self-interest is inextricably mixed with concern for the whole ecosystem, in contrast with the Midwest, where the prairies provided an incentive to restore an ecosystem that was generally understood to have little economic value, at least in the short run. Apostol, who grew up in the Chicago area and first encountered restoration on a visit to Bob Betz's prairie restoration project at Fermilab, suggests that this helps explain why restoration of ecosystems such as savannas and grasslands, sagebrush steppe, and even old-growth forest, which cannot compete with younger, more intensively managed forests in economic terms, came into the rich mix of restoration efforts in the Pacific Northwest rather late. "No one eats owls," he points out.[22]

In their book, Apostol and Sinclair note that the link between restoration and natural capital "makes some people uneasy because it introduces economic valuation into what many think should be a purely altruistic pursuit."[23] This implies a conception of ecocentric restoration even more stringent than the one we have adopted in this book. Yet the conservation culture of the region now supports numerous projects that exemplify this ideal. For example, Steve Moddemeyer, who was for a time senior strategic advisor in the director's office of Seattle Public Utilities, points to the Cedar River Habitat Conservation Plan, finalized in 2002 to protect the watershed that provides water for the city of Seattle. Here the city has acquired an entire watershed—some 36,000 hectares of land—that provides two-thirds of the water for the city and has turned it into a preserve, but one that allows for the ongoing restoration needed to preserve an actual ecosystem rather than just a hydrological resource. Driven in part by the need for a reliable supply of clean water but also by requirements to improve habitat for the endangered chinook salmon under the Endangered Species Act, the project has gone a step beyond utility and compliance to include a more comprehensive—indeed, ecocentric—objective. Not only has this added to the cost of the project, it has also meant constraints on human use of the area. Casual visitors are excluded, and only those working on water supply operations or restoration are admitted. This includes Native Americans pursuing traditional activities such as gathering plants for food, medicine, and ritual—uses that are in themselves restorative since these activities helped shape the precontact ecosystems that are being restored.[24]

Indigenous Peoples

The tribal peoples of the Pacific Northwest have not been alone among Native Americans and indigenous peoples generally in undertaking restoration initiatives as part of a program of environmental and cultural recovery. Dennis Martinez, of mixed Tohono O'odham, Chicano, and Swedish heritage, whose work as chair of the Society for Ecological Restoration's Indigenous Peoples' Restoration Network since its founding in 1995 gives him an international perspective on these developments, notes that many of the 564 recognized tribes in the United States have begun such projects in recent years.[25]

In addition to the immediate benefits they offer those involved, such initiatives offer a valuable perspective on restoration and its various meanings. They are in an important sense the ultimate form of restoration—that is, restoration of the whole ecosystem, including its human inhabitants, often engaged in the cultural activities of hunting, fishing, gathering, firing of vegetation, and agriculture that originally shaped the ecosystems involved. At the same time, this changes the valence of this work from engagement with an ecosystem that owes nothing to the inhabitants to what Fikret Berkes characterizes as "livelihood" and the maintenance of their habitat. Recognizing this, leaders in this movement have given it a name—*biocultural restoration*—to refer to an idea that stands in an interesting relation to the idea of ecocentric restoration.

Reflecting a worldwide trend on behalf of recovery of traditional cultures, these projects typically entail a complex mix of cultural and economic challenges, not least of which is the challenge of sustaining an economy in a capitalist system. Martinez says that indigenous peoples seeking to revive a place-based culture typically struggle to balance traditional land use practices with economic development. In doing so, some have set long-term economic goals that benefit from restoration of the historic ecosystem while ensuring the long-term survival of culturally important plants and animals that may have little or no market value but are valued as kin or have value in a mixed-subsistence economy.

As we have seen, these two considerations come closest to coinciding in ecosystems such as the forests and fisheries of the Pacific Northwest, which support economies based in part on activities such as commercial fishing and recreational hunting and angling, which are modern versions of the technologies that shaped these ecosystems in the first place. It is when economics and the ecology of the old ecosystems *don't* overlap that we find the clearest examples of ecocentric restoration. Martinez notes, for example, that there is disagreement between traditional and progressive elements of the White Mountain and San Carlos Apache in Arizona over the reintroduction of wolves. He points out that whereas the "traditionals" favor reintroduction, seeing wolves as part of creation and therefore sacred, a few Apache ranchers do as well, and it is in the position of these ranchers, who support recovery of wolves at some cost to their own interests, that the disinterested, ecocentric motive is most evident.

Martinez notes that such projects reflect the values of traditional cultures and their intimate relationship with the ecosystems they inhabit. This is obviously quite different from the newcomer's notion of re-creating conditions that prevailed at the time of cultural contact, and that may not be the kind of habitat he or she prefers to inhabit. Although Martinez notes that interest in historic ecosystems has grown in recent years, rarely, if ever, are attempts to restore them motivated by concern for the old ecosystem for its own sake. Dave Tomblin, who has studied the development of restoration by the White Mountain Apache in Arizona, notes that the Apache are interested less in the re-creation of historic ecosystems than in recovering a vital relationship with the land while finding a way to survive economically in a Western, capitalist society.

Indeed, what Martinez has dubbed "kincentric," referring to the experience of other species as members of a family to which one belongs, is not the same as "ecocentric," an idea that reflects the experience of nature as *unfamiliar*. This makes little sense for a people who, as Martinez says, make "no distinction between humans and an environment that is *out there."* For them, as ethnobotanist Kat Anderson has written, "the 'New World' is in fact a very old world."[26] It is their habitat, and their management of it is naturally some version of what Marcus Hall calls "maintenance gardening." Such a culture may foster respect for all species. It may cultivate a stewardship ethic based on a sense of nature as sacred. But it has little use for the idea of *ecocentric* restoration, which is a response to the experience of an ecosystem as other than us, here before we got here, and so not our habitat.

This being the case, restoration projects initiated by indigenous peoples are typically examples not of ecocentric restoration but of adaptive management. The Menominee, for example, have managed their land in northern Wisconsin for sustainability for nearly a century and a half,[27] and the results are now visible from outer space, a dark rectangle of forest on a background of cleared land. Economically, the forest, managed on a sustained-yield basis, provides a steady supply of jobs and income. Ecologically, the results are clearly positive but complex, at least as far as the historic ecosystem is concerned. Donald Waller, an ecologist at the University of Wisconsin–Madison who has written extensively on the forests of the region, describes them as "hard to categorize" in historical terms. Consistent with Martinez's observation that such projects rarely reflect an interest in the historic system *as* a historic ecosystem, Waller notes that the Menominee have not attempted actually to restore the old forests and that their "low-impact management" has resulted in a system different from the old system, though resembling it in some ways. In particular—and of special importance to us—he notes that this has entailed integration of forest and wildlife management, notably "welcoming back wolves and retaining deer at far lower (and historically appropriate) densities than most of the rest of northern Wisconsin," where wolf recovery efforts have been controversial and management of deer herds reflects the interests of sport hunters.[28]

Martinez offers a similar account of a major restoration effort being carried out by the Santa Clara Pueblo in New Mexico. After the devastating Cerro Grande fire that burned more than 18,800 hectares of tribal land in 2000, the pueblo took over management of the burned-over land from the Bureau of Indian Affairs and closed Santa Clara Canyon to tourism in order to restore the forest, an effort that has so far entailed planting some 1.7 million ponderosa pine, Douglas fir, blue spruce, Engelmann spruce, and white fir seedlings. It has also embarked on a comprehensive invasive plant removal program and is restoring streamside vegetation and beaver.

All this makes this a striking example of what Mike Rosenzweig would recognize as reconciliation ecology, broadly overlapping with but not quite
coinciding with the notion of ecocentric restoration.

While drawing on traditional forms of land management,[29] such efforts
often draw on western techniques, knowledge, and ideas as well. Dave
Tomblin, for example, notes that restoration efforts by the White Mountain Apache owe a great deal to the influence of the Indian Division of the
Civilian Conservation Corps, which was active on the reservation in the
1930s.[30] Martinez says that his experience as a diplomat working with both
knowledge systems has led him to think of them as complementary. He
notes that what an indigenous culture typically brings to restoration is both
a conception of "spiritual reciprocity" in relationships with the environment and a repertory of actual land use practices based on traditional ecological knowledge. He notes that indigenous peoples often draw on ecology and other natural sciences for information related to changes in
migratory routes, oceanic currents, genetic viability of populations, disease
vector tracking, and large-scale biogeochemical and climactic shifts.[31] He
also points out that scientists have often benefited from the ground-truth
knowledge of place-based peoples. The Inuit and Inupiat, for example,
noted the thinning of Arctic sea ice in the 1960s, a development not confirmed by researchers using passive microwave technology until 1979. And
indigenous estimates of the populations of culturally important species
such as bowhead whale and caribou have consistently been more accurate
than estimates by western wildlife and marine or fisheries biologists.[32]

When this syncretism is successful, the result is a true cultural hybrid,
a land management program that brings together essential wisdom from
two cultures that have often been cast as incompatible. And this, Martinez
notes, is often reflected in collaboration, with tribes working in partnership with government agencies and institutions of higher learning. Increasingly, this entails the return of tribal peoples to publicly owned lands
from which they were evicted during the era of settlement.[33] For example,
Martinez notes that the Karuk of Northern California are reintroducing
fire, taking out roads, and restoring slopes under memoranda of understanding with the Klamath and Six Rivers National Forests and agreements with federal agencies in the Mid-Klamath Watershed Council.
Similarly, the Timbishe Shoshone now have access to some cultural resources in Death Valley National Monument and are comanaging some
areas with the National Park Service.

Of course, discrepancies remain. Martinez notes that his insistence that
indigenous peoples are often a keystone element, with small populations of

people playing a disproportionate role in the ecology of the ecosystems they inhabit, has resulted in disagreements with preservation-oriented environmentalists such as David Brower and conservation biologists such as Reed Noss. He thinks that ideas of ecological integrity promulgated by both conservation biologists and the Society for Ecological Restoration, with their emphasis on the resilience and self-sustainability of natural ecosystems, downplay the role of the people who in many cases both shaped and maintained the ecosystems being restored. He points out that restoration of these ecosystems entails ongoing, intergenerational maintenance and renewal. Burning is an emblematic example because fire-adapted ecosystems, in many cases created and maintained by a history of anthropogenic burns, are often more stable than so-called climax ecosystems.

Far from seeing this as a nuisance or liability, however, Martinez sees the human role in human-shaped ecosystems as a key factor in whatever ecological resilience they exhibit in response to developments such as altered land use patterns or global climate change. A culture that is integrated into its habitat has, Martinez points out, a powerful incentive to maintain that habitat, because the culture depends on the ecosystem in many ways. And to the extent this is true, such a culture is itself a crucial form of natural capital, not only maintaining biodiversity but also lending the ecosystem a resilience and adaptability it might not otherwise have.

This means acknowledging that the old ecosystem reflects a history of human influence, and so is to that extent dependent on activities such as burning, selective harvesting, fallowing, horticulture, creation of habitat reserves, and hunting and fishing. Of course, as Martinez notes, cultures no less than ecosystems are dynamic, and when these practices change, the ecosystem will change, too, as in the case of the Menominees' carefully managed forest. It is at this point that the disjunct with ecocentric restoration becomes evident, and the old ecosystem will be kept on its historic trajectory only if the old conditions are maintained and technologies carried on out of sheer respect for the old ecosystem. This entails the kind of "Sabbath" exercise exemplified by the restoration of prairies in the Chicago suburbs or the Apache ranchers who defer to the wolves. Here the old ecosystem is not regarded as "our" habitat, at least in an economic—or "livelihood"—sense but as an ecosystem to be maintained, preserved, and respected primarily for its own sake. Making a replica of an old thing is very different from having made it in the first place or maintaining it, and this is as true of an ecosystem as it is of, say, a classic car.

As we noted at the outset, ritual plays a crucial role in the creation and maintenance of relationships, and awareness of this may be one of the most important—and least recognized—contributions indigenous and premodern peoples have to make to conservation practice. Ritual and ceremony are integral to the kind of ecosystem-based adaptation that characterizes successful place-based cultures, reflecting the realization that humans play an essential role in maintaining the natural order.[34] Martinez notes that some tribes have revived ritual traditions abandoned during the period of displacement. He himself has participated in the revival by the Takelma Intertribal Project of the Salmon Homecoming Ceremony on the Applegate River in southwestern Oregon. Others have created new rituals in conjunction with their restoration efforts. For example, the Klamath Tribes of south-central Oregon have created a Sucker Fish Ceremony for that endangered fish. Burns, of course, are not only an important land management technology, but also a dramatic symbol of death and renewal, and although they have forgotten the songs and ceremonies formerly associated with prescribed burns, some tribes in California and Oregon still regard a burn as a spiritual event and offer prayers for a successful burn.

Restoration efforts of this kind, bringing back both cultural and ecological elements, give such efforts value beyond ecology as a nexus of cultural interaction. This may prove crucial. "If ever there was a need for equitable and reciprocal exchange of information and expertise," writes Thom Alcoze of Northern Arizona University, who works on ecocultural restoration with Paiute communities in northern Arizona, "it is in the context of ecological restoration on Native American Reservations."[35]

The Southwest

Of course, even when restoration of a historic ecosystem coincides with the interests of society, broadly conceived, there is plenty to argue about. In the Southwest, for example, researchers have compiled a comprehensive picture of the region's historic forests. They have determined that the ponderosa pine forests that occupied large areas in the region were generally open and parklike before European settlement, with densities of mature trees in the range of 50 to 150 per hectare. This savanna-like condition resulted from a combination of factors, including an arid climate and frequent, light surface fires, often ignited by the ecosystem's human inhabitants. It changed dramatically as Euro-Americans settled the area, suppressing fire and establishing an economy based on livestock grazing, industrial timber harvesting, and recreation.[36] Under these conditions, trees proliferated, creating fire-prone dog-hair thickets with tree densities as much as ten times those that prevailed before settlement.

Forest restoration in the region today entails reintroducing fire, thinning post-settlement trees, and removing heavy fuel loads to move the vegetation back toward the conditions that prevailed in presettlement times.[37] Because this drastically reduces the intensity of fires, it also increases the value of the ecosystem as human habitat, especially in the urban–forest fringe. Despite this, however, a plan, based on research by scientists at Northern Arizona University's Ecological Restoration Institute, to reinstate the historic fire regime has been beset by the conflicts common to any undertaking that involves the management of public lands. As of 2006, after a full decade of effort, the Greater Flagstaff Forest Partnership had accomplished only a third of the restoration projected in 1996. The Blue Ridge Demonstration Project in eastern Arizona treated only a fraction of the acres proposed for restoration, and in New Mexico, a number of projects have been scaled back.

An analysis of socioeconomic barriers to ecological restoration suggests that three major factors are involved. The first is insufficient funding. The second is the complex cultural history of the Southwest. Because ecological restoration raises basic questions about the relationship between people and their environment, it offers fertile ground for conflict. The third factor is subtle: the difficulty of precisely identifying the social, economic, and ecological benefits that accompany restoration of ponderosa forests. These include the income derived from harvesting small-diameter wood and biomass, improvement of soil and water quality, and enhancement of recreational opportunities — considerations that managers and officials often overlook in devising land management plans.[38] Just as ecologists have learned that the old ecosystems are not ecologically privileged, social scientists have learned that they are not socially or politically privileged, and plans to restore an area to its "original" condition can result in serious disagreement, no less so in rural or wilderness areas than in Chicago or San Francisco, as we have seen and will see again in the chapters that follow.

Chapter 8

Realization III: Finding a Voice

What we are seeing here is the emergence—that is, the discovery—of eco-centric restoration and the realization of its value as a conservation strategy. To the extent that this form of restoration challenges both the utilitarian emphasis of classic conservationism and the hands-off preservationism that characterized the environmentalism of the 1960s and 1970s, this constitutes something of a revolution in environmental thinking and practice. And although, as we will see, this has entailed certain intellectual, psychological, and political tensions that are still unresolved, it has been for the most part a quiet revolution and a productive one, in which restoration has gained general acceptance as a conservation strategy, enriching conservation practice while providing preservationists with a means of achieving their objectives.

The essential realization here has been that the sort of intensive management that went on at sites such as the Holden and UW–Madison arboreta in an attempt to reassemble an entire ecological system from the ground up differed only in degree from what has to be done to keep a "real," "natural" ecosystem, such as a fire-dependent meadow or savanna in a national park, on its historic trajectory in the face of novel influences from outside the system.

This is important because it has to do with how managers think and talked about what they were doing, which is crucial to the process of discovery and realization. Small-scale, intensive projects such as the prairie restoration projects at the UW–Madison Arboretum were uncontroversial: Who would object to an attempt, however quixotic, to turn an abandoned pasture into even a crude representation of a tallgrass prairie? At the very least, you could hope to get a few native species back into the landscape, and the only loss would be an old field rapidly turning into a thicket of weedy trees. But areas perceived—indeed sacralized—as natural or wilderness are a different matter. In such places it is possible, if you squint (or don't really know what you are looking at and are not familiar with its history), to sustain the illusion of pristine nature untouched and unsullied by humans. In fact, that is exactly what the National Park Service did for roughly a century and The Nature Conservancy did on its holdings for several decades. Besides this, attempts to restore an ecosystem by reintroducing extirpated species or attempting to control invading exotic species often fail. And even when they largely succeed, they challenge the illusion of unspoiled nature. Restorationist Steve Packard recalls how this affected his own thinking when he began working with the remnant prairies and woodlands in Chicago's Forest Preserves in the late 1970s. He and his colleagues confined their activities to obviously degraded remnants because, he says, "I—we all—felt unworthy" to extend the work into adjacent, higher-quality remnants, where the policy was to respond only to activities such as dumping that were regarded as intrusive"[1]—as though three-quarters of a century of protection from fire wasn't ecologically more disruptive than a bit of dumping.

This being the case, something important happened when land managers working with areas designated as natural, whether under the rubrics of wilderness, wildland parks, public hunting grounds, or nature preserves, began to realize that the associations in their charge were not really preserved at all. In fact, they were drifting ecologically in response to both internal and external forces, both losing species and picking up exotics—more often than not aggressive, weedy species—and in the process losing native biodiversity and changing their character. Of course this was precisely what George Wright and his colleagues had discovered and pointed out in the national parks in the 1930s, the inevitable result of looking at the ecosystem objectively, in ecological and historical terms. This raised a whole series of troublesome questions that managers had been able to ignore as long as they were working under the rubric of preservation. Over the years, however, attentive managers had gained a deepening sense of the dynamic character of ecosystems, of the impossibility of insulating them effectively from outside influences and of the need for active management to compensate for these influences, all of which led them inexorably toward the idea of restoration.

The Word

An early articulation of this thinking was the publication, in 1977, of *Recovery and Restoration of Damaged Ecosystems.*[2] Edited by John Cairns Jr. and several of his colleagues, this volume explored questions arising from human dependence on nature and urged the development of restoration as a discipline that would reflect a "future primitive" philosophy and a sound understanding of regional ecologies and their limitations and vulnerabilities. Three years later, ecologists Peter White and Susan Bratton summed up this growing awareness of the role restoration had to play in preservation in an article in which they pointed out that "it is impossible to remove human influences from reserves." They also argued that "active management . . . is necessary if only because of the permeating human influence," and they specified a whole series of measures, including "re-establishment" of natural and historic communities, "re-creation or maintenance of anthropogenic communities," "re-introduction of native species," and "control of animal populations that may be out of balance" as elements of policy for lands intended to "preserve historic conditions,"[3] that add up to a prescription for ecocentric restoration.

What was missing was the word, which White and Bratton did not use in their article. That same year, however, John Cairns organized a conference on the recovery process in damaged ecosystems, at which the idea of restoration finally hit home among ecologists. In the lead talk, ecologist Robert McIntosh focused on the relationship between restoration and succession and attacked the distinction between natural and human-caused disturbances of ecological systems, arguing that, as far as the ecosystems were concerned, the distinction was meaningless. The session, Cairns recalls, "was jammed, which startled me, and I realized that mainstream ecologists were beginning to pay attention."[4]

By this time, hundreds of people were doing projects that more or less fit our definition of ecocentric restoration, and *restoration* was finding its way into the land management vocabulary. This was something new. Lots of people were doing restoration at the time, but, odd as it may seem, there was no commonly recognized word for what they were doing. Bill Jordan recalls that when he and his colleagues were recruiting articles for the early issues of *Restoration & Management Notes* they often had to explain to contributors that what they were doing was indeed something that people were beginning to call restoration and that an account of their work would make a suitable contribution to the new journal.

Application of this old word to this in some ways new thing, obvious as

it may seem, took some time. Although the word had been applied in conservation contexts at least since Marsh, that was different—that is, the
thing being restored was different. Then it had been a resource or a quality
that was being restored; now it was the whole ecosystem, which had a very
different feeling. In any event, this was a new application of the word and
an important step in the realization of the new idea. This sort of thing is
hard to document; after all, who other than an editor scrambling for copy
was keeping track or even paying attention? But a few anecdotes, reflecting
the experience of practitioners who have played conspicuous roles in the
realization of ecocentric restoration, reinforce the point.

Robin Lewis III, reflecting on his work setting up seagrass restoration
experiments in Tampa Bay in the early 1970s, recalls that "no one was talking about restoration at the time, and I didn't think about what I was doing
in those terms. I just knew the mangroves and seagrass beds were declining, and I was interested in finding out whether they could be recovered in
some way." Florida restorationist Andy Clewell told us a similar story. "I
was doing restoration down in Florida and no one else was," he recalled.
"And then I ran into a copy of *R&MN* and I was amazed. I thought to myself, hey! people are doing this stuff up there in the Midwest, too." Tein
McDonald recalls colleagues in Australia copying issues of the young journal and passing them around. And as late as 1987, University of Pennsylvania ecologist Daniel Janzen, who works mostly in Costa Rica and was, by
his own admission, a bit behind the linguistic curve, described how acquisition of the word *restoration* had given definition and identity to a project
he had launched to restore tropical dry forest in Costa Rica's Guanacaste
National Park a few years earlier—a story that evokes the story of Helen
Keller suddenly grasping the word *water* in her family's pumphouse.

"See, I wasn't thinking about restoration," Janzen recalled. "I didn't
know anything about all the thinking about restoration that had been going on . . . up here in the States and other places. I didn't know anything
about that literature. . . . What was clear to me was that if I wanted dry forest—which I did—I was going to have to grow it back. I didn't even know
what to call this process, and for a while, before I heard about restoration
and restoration ecology, I was using the English word 'reconstruction.'"[5]

This is a striking fact. Half a century after Aldo Leopold had described
the plan for the restoration project at the UW–Madison Arboretum as
"something new and different," deeply engaged, well-informed people
were still in a sense inventing restoration on their own, feeling that it was
something of a novelty and groping for a vocabulary with which to talk
about it.

What this shows quite clearly is that, although some conflated this form of land management with the restorative land management practices that had been part of humans' relationship with their environment for millennia, those directly involved considered—and presumably experienced—what they were doing as something quite different. Clearly, ecocentric restoration, though acted out in early projects such as those in Ohio, Wisconsin, and Australia, was still a new *idea* nearly a half a century later, had yet to be realized—that is, made real—and in this crucial sense had not yet happened. Like the Viking or Chinese explorers who happened on a continent long before Columbus, the early tinkerers and experimenters had failed to recognize or articulate the importance, the distinctive value, and the promise of what they were doing. That, as we are seeing, has been the work of a later generation.

A Separate Development

Despite the obvious similarities and overlaps between disciplines such as forestry, soil science, limnology, and range management, and of more recently formed disciplines such as conservation biology, landscape ecology, and even land reclamation, sociologically speaking the relationship between restoration in the tradition we are dealing with here and these ancestral and sibling disciplines has been a distant one. This is especially striking considering that all these could be called restorative or healing disciplines. Indeed, those involved in conservation efforts in the mid-twentieth century often used variants of the word *restoration* to describe what they were doing. A scan of the *Journal of Forestry*, for example, picks it up in a scattering of articles dating back to 1921. And, in 1965, responding to the passage of the Wilderness Act the previous year, forest ecologist M. L. Heinsel, underscoring both the relevance of restoration to forestry and the disjunct between the two, noted that the commitment to the preservation of "wilderness areas and primitive parks" called for "a new and positive approach," involving "maintenance, or where necessary, restoration of natural forest communities." Identifying this as "a new field of resource management" that is "still virtually untouched by our profession," he noted that many foresters see it as "a negation of what we stand for" but argued that it offered both a challenge and an opportunity for a land management profession that "can sense values in a landscape that transcend the stumpage value of the trees growing upon it," and he concluded, "I hope it will be our profession."[6] As the reception of restoration by organizations such as the National Park Service and The Nature

Conservancy illustrates, however, the institutionalization and professionalization of restoration proved complicated.

Something new was taking shape as those involved began to distinguish between restoration and conservation protocols that might be described as restorative. This is why practitioners such as Dan Janzen and
Robin Lewis experienced the discovery of the word *restoration* as a label
for what they were doing as a small revelation. A word that had been in use
for generations in allied professions eluded practitioners of the new craft
and then took on a new meaning when applied to their work. Consistent
with this, it is notable that the restoration community represented by, say,
the Society for Ecological Restoration (SER) has developed pretty much
independently of these other professions. This is notably the case with respect to land reclamation, as represented by organizations such as the
American Society of Mining and Reclamation and the Canadian Land
Reclamation Association. Despite the obvious relevance of the techniques, if not always the aims, of land reclamation to the practice of ecocentric restoration, the two have developed as separate, rarely communicating cultures. Similarly, although limnologists were involved in lake
restoration efforts by the 1960s, the culture of lake restoration and the restoration culture represented by SER have remained mostly separate.
Other disciplines have stayed in touch. Landscape architecture, which
played a key role in the invention of ecocentric restoration, has continued
to play a supporting role in its development. And there has been much interaction with forestry and wildlife management, the two disciplines that
Aldo Leopold brought together in his own early contributions to the invention of ecocentric restoration. Indeed, in a way that M. L. Heinsel
would surely find gratifying, foresters participate in restoration conferences, contribute to publications such as *Ecological Restoration* and *Restoration Ecology*, and have developed practices strongly informed by the
idea of ecocentric restoration.[7]

At the same time, there has always been a sense that when this happens
a conceptual, psychological, and cultural boundary is being crossed.
When, for example, old-timer conservationists who showed up at the early
SER conferences mentioned that they had been doing restoration for decades, as foresters or game managers, the comment found little resonance.
No doubt this reflected the interest of a new generation in having its own
thing, being involved in the invention of a new way of managing land and
relating to nature. But it also reflected a sense that there was a real difference between, say, sustained-yield forestry and land management aimed at

or at least inspired by the notion of summoning back an historic ecosystem or landscape, not only as a resource but primarily for its own sake.

Indeed, the tension between these two aims became quite clear when, in 1989, just as SER was taking shape as an organization, the director of the California Foresters Registration Office of the State Board of Forestry pressed for a reinterpretation of a 1972 law that would have extended the authority of professional foresters over "wildland forests" to include all natural lands or wildlands as well. Not surprisingly, restorationists saw this as a kind of professional land grab by old-guard foresters. They also realized that it would mean that restorationists and all natural resource professional working on wildlands would have to be certified foresters and, thinking that that qualification had little relevance to the kind of work they were doing, they protested. Acting under the aegis of SER, they asserted the distinctive nature of their work and questioned the competence of a traditionally trained forester to carry it out. "Out of all the Registered Professional Foresters in the State, how many would allege to be qualified to perform all types of restoration work, including forest and non-forest vegetation types?," California restorationist Marylee Guinon, who led the protest on behalf of the newly constituted profession of restoration, wrote to board chairman Hal Walt that summer. "How many . . . have actually planned or implemented a restoration project in California?"

Eventually, the extension of authority was denied when, with the support of key figures such as Henry Vaux, dean emeritus of forestry at UC–Berkeley and author of the 1972 State Forest Practices Act, and Doug Leisz, associate chief of the Forest Service under the Carter administration, the state legislature approved a bill, sponsored by a number of professional organizations, including the California chapter of SER, that halted the board's "regulatory expansion into licensing of biologists."[8]

Of course, this is the obverse of how the National Park Service had responded to the restorationists in their midst a half a century earlier. Together, the two stories illustrate two ways an institution or a profession responds to something new: either by expelling it or by trying to appropriate it.

A friendlier reflection of the emergence of restoration as an item in conservation circles at this time came from *Land and Water*, an environmentally oriented trade journal for contractors and engineers that in 1992 announced that it was "repositioning" into the "natural resource and management market" and added the line "The Magazine of Natural Resource Management and Restoration" to its masthead.[9]

Articulation

What was crucial here, as in any invention or discovery, was the realization of the distinctive value of the new thing or idea and the articulation of that value. Something people had been doing for the better part of a century was acquiring a name and with it an identity and a measure of legitimacy. Managers who had been walking the walk, in some cases for years, now began to talk the talk, and as a result what they had been doing acquired not only a label but also a voice in the conversation about natural area conservation.

At the same time, they acquired an answer to the question Aldo Leopold had left hanging when, in the 1940s, he began to blur the line between beauty and utility with regard to the natural landscape, arguing that the most complex biota is not only the most beautiful but also the most useful.[10] This is obviously an appealing idea. But it leaves open the question of what would become of an ecosystem if its human inhabitants did not happen to find it either useful or beautiful.

This question of the fine line between beauty and utility—or between ecocentric and self-interested land management—is fundamental. In fact, it is one of the great questions, and uncertainty about how to handle this and talk about it has both delayed recognition of ecocentric restoration as a distinctive form of land management and complicated thinking about it. At the UW–Madison Arboretum, for example, when in the late 1970s and early 1980s Bill Jordan proposed several projects related to research, education, and public outreach that he felt built on the arboretum's early restoration efforts, there was little interest and even a good deal of resistance to the idea of making a big deal out of what some had come to think of as just a kind of ecological housekeeping or "glorified gardening". Eventually, several of these projects were carried out, however, and by the late 1980s they had become an integral part not only of the arboretum's program but of its reputation and institutional identity.

Two Journals and a Society

One of these projects, the journal *Restoration & Management Notes* (since renamed *Ecological Restoration*) turned out to be one of two journals that played a role in the ongoing discovery of restoration that reached a kind of threshold around this time. The first, by a few months, was the *Natural Areas Journal*, published by the recently formed Natural Areas Association, the first issue of which appeared in January 1981. The first issue of *R&MN*

In the Beginning . . .

As the Gospel of John (1:1) suggests, the naming of a thing is the crucial act that realizes it—that is, makes it real—and it seems that even the term *ecological restoration* fits this pattern, appearing in conversation long after the earliest restoration projects were launched but early in the process of realization. Prompted by a question from Karen Rodriguez, in the Great Lakes office of the Environmental Protection Agency, we asked Frank Cook, who is writing a history of the UW–Madison Arboretum, when he first encountered this term in the arboretum's records. He said that he did not see it at all before the 1980s, although Jim and Elizabeth Zimmerman, naturalists who had close ties to the arboretum for many decades beginning in the 1950s, used the term *ecosystem restoration* in describing what they characterized as the arboretum's distinctive work in a 1973 newspaper article. Interestingly, Cook did not find the more general term *ecological restoration* anywhere in the arboretum's archives until 1984, when it appeared in an article in *Restoration & Management Notes.*[a]

---

a. Frank Cook, personal communication, May 11, 2010. The Zimmermans' article appeared in *The Wisconsin State Journal*, January 21, 1973. We located the first use of the term *ecological restoration* in *R&MN* by word-searching back issues of the journal electronically. It appears in the introduction to an interview with John Berger, Volume 2, No. 2 (1984), p. 68.

---

appeared just a few months later, in July. Both journals focused from the outset on the management or stewardship of historic ecosystems, usually referred to as "natural," "pre-settlement," or, in *NAJ*, "wilderness" areas. Both asserted from the first that there was a growing need for management of these areas and pointed toward growing interest in this work and the need for both a new discipline to carry it out and better communication between those involved. Their virtually simultaneous appearance, which was entirely coincidental, attested to an emerging sense among land managers that managing ecosystems by leaving them alone was not a satisfactory strategy for ensuring their well-being or survival and that their management posed a challenge that existing disciplines were not addressing effectively.

In an article in the second issue of *NAJ*, ecologist Peter White asked how we could ensure that our natural areas and parks functioned to preserve species and natural systems. He followed this up with a discussion of challenges to preservation that amounted to a checklist of concerns for the restorationist: invasion by exotic species, loss of species due to imported diseases such as chestnut blight, and questions associated with the use of fire and the reintroduction of extirpated species.[11]

Similarly, taking a parallel if not identical course, the lead editorial in the first issue of *R&MN* asserted that the journal would deal "only with the restoration and management of ecological communities for essentially scientific or esthetic purposes." Where the two journals differed was in the emphasis on key ideas such as "preservation," "management," and "restoration." While the writing in the early issues of *NAJ* generally emphasized preservation and management of wilderness and natural areas, admitting "restoration" as though through the back door, *R&MN* placed it literally up front, in its title, focused on this form of management from the outset, and dealt with wilderness only insofar as it pertained to restoration.

From a purely practical point of view, this may seem a trivial difference. In fact, both journals dealt with restoration from the first, and many authors who contributed to one also contributed to the other. The difference is important, however, when we are considering the discovery of restoration and the realization of its distinctive value as a conservation strategy. And here *R&MN* took the lead in a way that reflects the different environments that gave rise to the two journals. *NAJ* was founded by a group of land managers who had come together to form an organization and who, as the name of the organization and its journal makes clear, had questions related to the preservation of natural areas very much on their minds. *R&MN*, in contrast, was dreamed up by just two people, Jordan and arboretum ranger Keith Wendt, whose office on the outskirts of Madison looked out over what they were coming to think of as the world's first restored prairie.

In retrospect, it seems clear that what we had here were two cultures converging on an idea: on one hand a culture of wildland managers and aficionados who tended to think in terms of animals and of ecosystems as animal habitat and on the other a culture of gardeners and botanists working in urban and suburban contexts for park systems, educational institutions, and departments of natural resources and transportation, who had little incentive to think in terms of wilderness and who placed more emphasis on plants than on animals.

The first of these cultures reflected the ambivalence regarding restoration and management of natural areas characteristic of the environmentalism of the 1960s and 1970s more clearly than did those who were working with ecosystems they regarded as disturbed and who were beginning to think of themselves as restorationists. Indeed, Jordan recalls a discussion of this matter at the Natural Areas Association's annual conference, held near Dayton in 1985. Having launched *Restoration & Management Notes* without one, he was feeling the need for a society to back it up. The arboretum management had backed the launching of the journal only grudgingly. It had no interest in the arboretum taking the lead in what they saw as an extension of that project, and he attended the conference in part to make a pitch for the NAA to adopt restoration as a prominent part of its mission. The participants welcomed the opportunity to discuss this matter, and at one point a dozen or so spent a couple of hours sitting under a big oak near the meeting hall talking it over. But in the end, they decided not to make—or at least to formalize—any such shift in emphasis. This is one reason why, three years later, Jordan was one of another group that came together to form the Society for Ecological Restoration.

Another event that helped brand restoration as a distinctive form of land management was the exuberantly titled "Restoring the Earth" conference held in Berkeley, California in January 1988. The organizer was John Berger, the young environmental journalist and ecologist who had profiled the work of Bob Betz, Ed Garbisch, and a handful of other restorationists a few years earlier in his book *Restoring the Earth*, itself an important early contribution to the task of articulating the idea and importance of restoration.[12] This was not the first conference to focus on restoration. The wetland creation and restoration conferences had been held at Hillsborough College annually since 1974; the North American Prairie conferences, which had been held biennially since 1968, included presentations on restoration of prairies; and a symposium organized to explore the idea of restoration ecology, a related but different idea, as we will see, was held in 1984. We believe, however, that the Berkeley conference was the first to focus on restoration itself rather than a particular ecosystem, emphasizing ecocentric aims and dealing with restoration in a comprehensive way, including considerations of planning, education, and philosophy as well as technique. Implicit here was the idea that restoration was a distinctive activity, not just a special effect or hobby but a protocol that applied to *all* kinds of ecosystems.

This was an important step in the crucial task of labeling restoration and identifying it as a distinctive item in the repertory of conservation strategies. Unlike the early Hillsborough conferences, which had only ambiguous institutional support, "Restoring the Earth" was cosponsored by the University of California's College of Natural Resources, its Center for

Environmental Design Research, and the San Francisco Bay Conservation and Development Commission. It turned out to be a major event, which included some 200 speakers and 1,200 participants from all over North America and a few from abroad. Moreover, the participants represented a wide range of disciplines, professions, and interests, providing perhaps the first concrete sense of the bench strength of the nascent community of people involved in restoration work. And it helped trigger a seminal National Research Council study of the scientific, technological, and policy aspects of aquatic restoration.[13]

A third event with far-reaching consequences for the development of restoration as a discipline was the creation, in the same year, of the Society for Ecological Restoration. This initiative grew out of a series of two conferences on "Native Plant Re-vegetation," organized by John Rieger, who worked for the California Department of Transportation. The first of these, held in San Diego in 1984, attracted more than 100 participants, and by the time Rieger got around to organizing the second conference three years later, the notion of restoration had gained currency to the point that the organizers began thinking the time was right to launch an association of some kind. With this in mind, Rieger recruited Bill Jordan, who, as editor of *R&MN*, brought into the conversation the prospect of a ready-made publication for such a group, and at the second "Reveg" conference, held in San Diego in April 1987, he and Jordan sat down with John Stanley and several other California restorationists and sketched a plan for the new organization. Thinking ahead, they decided to extend membership beyond California. Indeed, the first paid membership was from Bitterroot Nursery in Hamilton, Montana, and membership has since grown to 1,700 members from fifty-six countries.

During this period, leading advocates for the environment began drawing attention to restoration and touting its promise as the key to ensuring the survival of classic ecosystems through the upcoming century. "Here is the means to end the great extinction spasm," E. O. Wilson wrote in 1992, adding, "The next century will, I believe, be the era of restoration in ecology."[14] In an article on the emerging field, *The Chronicle of Higher Education* quoted ecologist Michael Soulé suggesting that conservation biology, the discipline he had helped define just a few years earlier, was a kind of end-game response to the challenge of saving surviving remnants of the old ecosystems, and that as these are protected, degraded, or lost entirely, "the job will have to be turned over to the restoration ecologists; . . . that will be the only thing left to do by the middle of the next century."[15] Bruce Babbitt, identifying restoration as the theme of his tenure as secretary of the interior under Bill Clinton, outlined a plan to create an entire national park virtually from scratch along the Missouri River between St. Louis and Kansas City to commemorate the Lewis and Clark expedition.[16] And David Brower set aside his initial resistance to the idea of restoration to champion the idea during the last years of his life. "Having been in the conservation business half a century, I should have thought harder about restoration sooner," he wrote in his autobiography (in a chapter titled "Restoration, a Blueprint for the Green Century"). "Unfortunately," he added, confirming our account of the neglect of restoration by two generations of conservationists, "many others have overlooked it too."[17]

Of course, what these boosters and converts had in mind was not always exactly what we would call ecocentric restoration, but it certainly included it and was obviously inspired by it and by the prospect for the future of the old ecosystems it implied. Brower, of course, spent a lifetime defending wilderness—a version of the "nature as given" that is the aim of ecocentric restoration. Babbitt unabashedly asserted the value of "going back in time" and made restoration of ecosystems in Everglades National Park, the first national park created as an ecosystem and not as a spectacular landscape, a keystone of his administration. And both Wilson and Soulé are notable bulldogs on behalf of biodiversity conservation, which they argue will depend to a great extent on what we are calling ecocentric restoration.

Chapter 9

Realization IV: Getting Real

In the meantime, what was happening on the ground? We have seen how, coming to The Nature Conservancy (TNC) in 1970, Bob Jenkins discovered an organization that had become adept at acquiring parcels of land but had almost no capacity for managing them. Seeing this as a weakness fatal to TNC's mission, he set out to remedy it.

"There was a plan of sorts," he recalls. "But as soon as I started working with it, I realized that it was just an exercise in theory because, whatever we had on paper, we had no management capacity at all. So I was really working from the ground up. I spent some time thinking about our aims. I decided that we would never have the area we would need to build carrying capacity in a serious way, but we could have enough to make a real contribution to species conservation and what we now call biodiversity. So I proposed that we concentrate on that—on what I called lifeboating for species and other ecosystem elements."

Realizing that implementing what he had in mind would entail a change not only in the policies but in the culture of TNC, Jenkins avoided the reaction that George Wright and his colleagues had experienced in the National Park Service (NPS) three decades earlier by, in effect, sneaking up on the organization like a runner tiptoeing up behind a rival before attempting a pass. "The advantage I had," he says, echoing Andy Clewell's comment about the benefits of getting in at the start with a minimum of regulation and oversight, "was that I was working by myself. Everyone else was working on acquisition, and they left me alone. So I wrote up the first long-term plan, and the organization just went along with it and helped raise money to support the projects I wanted to do. So long as we could get funding, no one interfered. It was a rare opportunity, really. The way it

161

worked out was that the whole management mission for the Conservancy came off my blackboard during the first few years I was on the staff."

Given TNC's aims, *management* would necessarily mean ecocentric restoration. Jenkins's early projects were what might be called mainte- nance restoration: attempts to control exotics and restore processes such as burning or hydrological cycles needed to return a system to its historic tra- jectory. But others involved intensive restoration of drastically altered lands on non-TNC property. What Jenkins recalls as "the first big one" be- gan one day in 1971 when a former physical chemist named Ed Garbisch walked into his office to propose large-scale restoration of tidal marsh around Chesapeake Bay. "I had thought about restoration as a possible way to expand preserves," Jenkins recalls. "I had no way to undertake such a project. But Ed had some ideas, and within a week we had a plan and had created the Center for Applied Research in Environmental Sciences (CARE) to carry it out." The next year, with funding mainly from himself and his family, Garbisch undertook his first large-scale project, bringing in bargeloads of silt to reconnect the two halves of Hambledon Island in the Bay and planting the resulting sand-flat suture with a quarter of a million native saltmarsh plants.[1]

Although Garbisch himself, who went on to pioneer the restoration business along the mid-Atlantic Coast, downplayed the idea of ecological authenticity and denied any interest in "going back in history,"[2] the project was catalytic for TNC, which of course *was* committed to ecological au- thenticity. A few years later, Jenkins worked with Bob Betz, another chemist-turned-restorationist, helping with negotiations with the Depart- ment of Energy that paved the way for the pathbreaking prairie restoration effort Betz proposed to undertake at the Department of Energy's Fermi National Laboratory in suburban Chicago.[3]

Although TNC wasn't directly involved in the Fermilab project, it played a catalytic role. And conceptually similar projects that Jenkins launched in the 1970s set TNC on course for what was to prove a major shift in the organization's acquisitions and land management policies over the next couple of decades. Complemented by initiatives in what has come to be called community-based conservation, this led to the transfor- mation of TNC's idea of its own mission and means for achieving it, and opened up a whole new dimension of value for restoration. This is an im- portant chapter in our story of discovery and realization of value, and we will consider it further in chapter 10.

Summing up, Jenkins notes that restoration has become an important strategy for TNC, which now routinely acquires degraded lands as part of a strategy to expand existing preserves or create corridors or ecological stepping-stones between them. However, much of this work proceeds "under the radar,"[4] he says, a fact he attributes to the habit of emphasizing business at the expense of ecological considerations that prevailed in TNC during the nearly three decades he was on the staff and also perhaps to a lingering concern that restoration compromises the naturalness of an ecosystem.[5]

This "under the radar" business is of great interest to us because we are concerned here not just with the implementation of restoration but also with the realization of its value, which depends on the way the work in the field is validated by the organizations involved and how those involved think and talk about what they are doing.

Realization of an activity entails not only doing it or even institutionalizing it. It also entails acknowledging it as part of an organization's—or a person's—mission and identity, talking about it, writing about it, even bragging about it—all developments that have taken place in TNC only recently.

The National Park Service Redux, Again

Turning to the NPS as our agency parallel to TNC, we find a similar pattern of rapid development over the past couple of decades, with acknowledgment and realization just a step or two behind.

As we have seen, the NPS had encountered the idea of ecocentric restoration on two prior occasions, first in the 1930s and again three decades later in the form of the Leopold Committee's report in 1963, but had resisted it both times. The second time, however, historian Richard Sellars found that the Leopold Report did initiate changes, including a gradual buildup of scientific expertise within the agency and more emphasis on science in defining policy. This took time, but it did set the agency on course for fundamental changes, and these finally began to take hold in a serious way in the 1990s, in concert with the acceptance of restoration as a key element in conservation practice within the conservation community generally. Although the NPS had not abandoned its commitment to preserving and showcasing spectacular scenery, it had been making a gradual transition to a management policy that included restoration. In 1988, the NPS policy manual emphasized the importance of restoring fire as a natural process, an acknowledgment especially important for the stands of giant sequoias in Yosemite, Kings Canyon, and Sequoia National Parks. A few years later it included in its "Vail Agenda" a brief acknowledgment that ecologically sound management "requires the maintenance or restoration of native ecosystems and resistance to the establishment of alien organisms." Today, management strategies focus on restoration of processes such as burning and on removal of abandoned or obsolete infrastructure, including asphalt roads, visitor facilities, and old tile fields, which alter the hydrology of meadows and wetlands.[6]

At Yosemite, for example, restoration ecologist Sue Beatty notes that the park has been conducting prescribed burns since the 1970s and over the past ten years has burned an average of 6,729 hectares a year, counting both lightning-caused and management-ignited fires. She also points to recent projects involving removal of dams, social trails, abandoned utilities, and ditches in meadows that are focused on restoration of hydrology to support natural processes. Other projects include control of invasive exotic plants in 16 hectares of Himalayan blackberry in Yosemite Valley and restoration of the 0.8-hectare Happy Isles fen, a rare example of a California fen ecosystem.[7]

Although projects of this kind are now fairly common in the national parks, those involved think there is plenty of room for more. Looking at the parks overall from his perspective as conservation and outdoor recreation chief with some thirty years' experience with the agency, Rick Potts expresses a mix of satisfaction and impatience. Acknowledging that his own shift from preoccupation with preservation of undisturbed wilderness to the realization that restoration has an important role to play in the kind of preservation implicit in the agency's mission dates back only to his encounter with the work of people such as Dan Janzen twenty or so years ago, he still thinks more could have been accomplished in that time. "Work like that at Yosemite has certainly been a step in the right direction," he says, "but I'm not sure we've fully realized that our job is not just inventorying what we have—not just diagnosing problems, but solving them." Acknowledging that drastic alterations in biota such as have occurred in Hawaii since European contact or those occurring in response to climate change may preclude restoration in the strictest sense, he suggests that this only increases the urgency of developing management plans that place restoration in the context of these efforts, and he expresses some frustration with the pace of progress in this direction. "Are we serious about restoration?," he asks. "And if we are, why don't we respond to problems such as the woolly adelgid or the spruce budworm the way we would respond to an unwanted wildfire? How are we prioritizing what needs to be restored? Are we stuck at the prospect of making the leap into the messy business of restoration?"[8]

Such impatience comes with the territory. Having assembled a plausible representation of prairie vegetation, for example, the ecocentric restorationist immediately begins to think about putting a few bison on it—and then perhaps a grizzly bear or two.

Bison

Bison and grizzlies provide a dramatic example of this logic of restoration at work. American bison, the largest quadruped on the continent at the time of European contact, obviously belong on the prairies—are indeed as iconic of the prairies as salmon are of the Pacific Coast. But whereas salmon spend most of their lives in the ocean, sharing habitat with humans only to spawn, bison spend *all* their time on the prairie, and they need a lot of it to make a living.

The challenge this poses for the restorationist is evident to everyone. Visitors to the UW–Madison Arboretum's prairies often ask, sometimes sardonically, "Where are the buffalo?" And although the early projects were obviously too small or too hedged in by city, suburb, or farmland to make such an addition possible, the prairie restoration project at Fermilab actually began with bison. The laboratory director acquired four animals in 1969—two years before Bob Betz showed up proposing to create a prairie inside the proton accelerator ring—to honor the prairie heritage of the site. Quartered in a fenced pasture, they represented just the first step on the restoration escalator. But as they undertook larger projects, managers, captivated by the idea of restoring the whole thing, found themselves wanting to go further.

In 1978 TNC manager Paul Bultsma set up a leasing arrangement allowing bison to graze on TNC's S. H. Ordway Jr. Memorial Prairie in north-central South Dakota. This worked out well enough that when Al Steuter and Bob Hamilton joined the staff shortly after Bultsma left, early in 1982, they thought it would be feasible to add bison to some TNC holdings as long as TNC owned the herd and managed it properly. They began pressing the Minneapolis office for approval of such a plan and, in 1984, with the support of managers Glenn Plumb and Mark Heitlinger, added eighteen bison to the biota of the Ordway Preserve. In keeping with TNC's insistence that the program be supported by a "grass endowment," the conservancy sells bison as calves and breeding stock and also participates in a flourishing market for bison meat, a good example of ecocentric restoration that also pays its way.

TNC now has some five thousand bison on 44,000 hectares in nine preserves in seven states from Kansas and the Dakotas to Colorado, so that at a few places, such as TNC's Tallgrass Prairie Preserve in Oklahoma, one *can* now see what several thousand acres of silphiums look like tickling the bellies of bison. But of course the logic of restoration presses on to the next question: What about the grizzly bears?

To see them, you have to go to Yellowstone National Park, where managers have been implementing a comprehensive program of reintroductions and restoration, which they now see as integral to the mission of the park as defined in the legislation that created it in 1872—that is, preservation of *all* its wildlife species. Interestingly for us, Glenn Plumb, who is now chief of aquatic and wildlife resources for the Park, views these efforts from the perspective provided by his early work with bison reintroductions for TNC.

Plumb, whose career has taken him from ranching and range management through work with TNC to his current work with restoration in perhaps its most ambitious form, characterizes this effort as representing the forefront of the development of restoration over the past two or three decades, and he notes that it reflects conditions and opportunities available hardly anywhere else in the lower forty-eight states. Key factors have been scale, long-term institutional support, and an economics in which the "product" of the restoration effort is independent of finances and not expected to pay its way.

Overall, then, restoration of this iconic species has progressed in the past three decades from a handful of animals displayed in captivity in the vicinity of a restored prairie to the development of the Yellowstone population. With the help of grizzly bears and the wolves that were reintroduced beginning in 1995, this population looks after itself—a rarely realized restoration fantasy.[9]

Plumb's point about finances goes directly to the distinction we are making between holistic and meliorative restoration. He notes that, in contrast with TNC's program, which depends partly on the "grass endowment" represented by the sale of bison, the program at Yellowstone is mandated by statute and funded by Congress and so is free of the economic constraints that may limit such efforts in almost any other context. He notes that although TNC's work has resulted in a highly successful business model for bison management, one that has been implemented on an even larger scale by Ted Turner, who manages bison on a for-profit basis on his vast holdings in the West, the effort in the national parks is fundamentally different and uncouples this work from any vestige of self-interest

---

Scaling Up

As restorationists have become increasingly comprehensive in defining the qualitative goals of projects, they have also made dramatic gains in scale. In the paradigmatic case of the tallgrass prairies, for example, the scale of operations has increased logarithmically, with practitioners adding a zero to the acreage of the largest projects under way roughly every decade for the past six decades, from 5 hectares at Green Oaks and the Morton Arboretum in the 1950s and early 1960s, to a planned 40,000 hectares now being restored at Neal Smith Wildlife Refuge in Iowa. This reflects not only the development of new techniques, such as direct seeding into existing sod[a] and the adoption of combines and mechanical seeding equipment to the demands of work with native species, but also a growing interest in the prospect of recovering this ecologically extinct ecosystem on an ecologically significant scale.

---

a. Steve Packard, "Successional Restoration: Thinking Like a Prairie," *R&MN* 12, no. 1 (Summer 1994): 32–39.

---

other than that associated with the value of free-ranging bison herds as an attraction for visitors.

Beyond Purism

Projects such as the bison reintroduction effort at Yellowstone are a long way from test-tube projects such as the early projects in which a handful of professors or aficionados tinkered with an ecosystem or bit of land, pursuing their notion of restoration in relative privacy. As practitioners take on larger projects, they inevitably find themselves working in places where people live. This raises the question of what relevance, if any, a project such as Henry Greene's prairie might have for the management of the environment of an entire region, such as the Great Plains, the Florida Everglades, or the entire binational watershed of the Great Lakes.

Here the logarithmic increase in the scale of projects we have noted over the past five decades loses its meaning, as the model provided by "purist" projects such as Greene's encounters realities that entail serious compromises.[10] We see this in a wide range of projects that extend some version of the notion of ecocentric restoration outside ecologically privileged areas such as the national parks or holdings of organizations such as TNC into places where large numbers of people actually live.

An excellent example is the work of the Environmental Protection Agency on behalf of restoration of the Great Lakes watershed.[11] From our perspective, what is especially interesting here is how the work of the Chicago restorationists influenced the movement of an agency, not in this case from environmental indifference toward environmental stewardship but from a narrowly focused form of remediation to an increasingly comprehensive program aimed at seriously large-scale restoration.

Karen Rodriguez, who has worked in the Great Lakes National Program Office (GLNPO) of the EPA since 1993 and has been a key player in the program throughout this transitional period, says that early on the agency focused mainly on abatement of pollutants from point sources under the Clean Water and Clean Air acts, Superfund, and other legislation. By 1990, however, much progress had been made in this area. Water quality was improving in many locations, and staff members were beginning to realize, as John Cairns had nearly a half a century earlier, that clean water isn't the whole story—that lakes full of clean water might turn out to be nothing but what then–GLNPO director Chris Grundler called "bathtubs"—and began thinking beyond water cleanup to actual restoration. Rodriguez herself was hired, she says, in part because of her experience as a volunteer with Chicago's North Branch restorationists. That experience actually had little to do with clean water, but the fact that the GLNPO saw it as an important credential suggests how the early initiatives in hard-core ecocentric restoration have influenced and even inspired more recent, "real-world" efforts. (Another of these was the creation of Chicago Wilderness, a consortium of several hundred organizations launched in the mid-1990s on behalf of biodiversity conservation in the Chicago region.) In 1992 Grundler and Russell Van Herrick, then with TNC, teamed up to create a Great Lakes office in TNC to focus specifically on preservation and restoration of biodiversity. Funded in part by the EPA, this was the first in a series of events in the 1990s that pushed the agency to adopt a program of restoration rather than just water cleanup for the Great Lakes.

The next was a workshop on savanna restoration, held at the EPA's Chicago offices in February 1993, and a follow-up conference that was attended by more than a thousand participants. Then the next year, the GLNPO took another step in the direction of incorporating ecocentric restoration into its repertory by funding TNC to write "The Conservation of Biodiversity in the Great Lakes Basin: Issues and Opportunities," a document that, Rodriguez thinks, "changed the way the Great Lakes community views the Great Lakes Basin." "[It] turned our office and EPA on its head," she comments. "Even though the Great Lakes Water Quality Agreement calls explicitly for restoration, we had been focusing narrowly on water quality, on deep-water areas of the lakes, and also on 43 'Areas of Concern' that had been identified back in the 1970s. That conveyed the impression that the Great Lakes were trashed and unrecoverable. But the TNC report not only identified more than 130 species and community types in the basin as globally rare, it analyzed ongoing threats to biodiversity and introduced the idea of restoration as a useful conservation tool."

Two years later, the GLNPO and Environment Canada cohosted the second "State of the Lakes Ecosystem" conference, held at Windsor, Ontario in 1996. "A number of people warned us not to bring up questions of land use," Rodriquez recalls. "The concern was that it would be too provocative. But we did it anyway—we even included a paper on land uses in the basin. And many of the participants were furious because they saw this as displacing the commitment to contaminant remediation that was the primary directive of the Agreement. But it worked out surprisingly well. And after that there was no question that both biodiversity and restoration were on the table to stay. Of course this wasn't altogether new. There was a lot of restoration going on around the country by 1996. But I think the context and the scope of this development was something new."

And productive. During the decade and a half since the Windsor conference, restoration has played an increasingly important role in conservation efforts on behalf of the Great Lakes, in the process attracting important top-down as well as bottom-up support, including President George W. Bush's Executive Order 13340, which led to the creation of the Great Lakes Regional Collaboration. It also paved the way for the Great Lakes Restoration Initiative by the Obama administration, which resulted in $475 million in funding for in-the-water and on-the-ground restoration projects. In addition, the GLNPO and other federal and state agencies, nongovernment organizations, academic institutions, and private companies are now sponsoring a wide range of projects, many of which include a strong ecocentric element. For example, Rodriguez points to removal of dams and placing of culverts to create habitat and allow passage of native fish, removal of invasive exotics such as phragmites now under way on much of the more than 217,000 hectares of coastal wetlands around the lakes, replanting of dunes with native marrom grasses, restoration of historic burn regimes on lakeplain prairies and savannas, and "softening" of engineered shorelines in key areas.[12]

All the Parts

Each of these agencies and organizations has provided a context for the development of ecocentric restoration, each with a different mission, responding to the idea of ecocentric restoration in its own way. Overall, challenged and perhaps inspired by the notion of re-creating the whole thing, practitioners have defined their objectives in increasingly comprehensive ways, moving their work toward the idea of restoring the whole system in all its aspects, including function and dynamics as well as composition and structure. On the prairies, for example, practitioners beginning with Ted Sperry had defined their goals principally in terms of the vegetation and made generous use of fire in large part because it favors native plants in their competition with exotic plant species in many ecosystems. A few, however, seeing that a prairie is as much an assemblage of insects as of plants, came to be concerned that too-frequent and too-thorough burns might be harmful to insects that spend the dormant season as eggs or pupae on the stems of plants, and began proposing more complex burn schedules that reflected this concern.[13] In a parallel vein, others have suggested that practitioners are relying too heavily on prescribed burns and underrepresenting the grazing that was a major factor in the ecology of the precontact prairies. This has engendered some debate over the notion of using more easily managed cattle as surrogates for the original antelope and bison for this purpose. Underscoring this point, Steve Packard recalls that some years ago, after a presentation by Al Steuter, someone objected to the idea of using surrogate species, saying they didn't want to see cows on a prairie. Steuter said simply and without heat, "We don't care"—a memorably blunt expression of the indifference to human preferences that lies at the heart of the idea of ecocentric restoration: We prefer bison, too, but if you can't have them, then put up with cows.

Overall, restorationists' concerns have expanded well beyond the early emphasis on plants to include animals, the most prominent—and naturally controversial—example of which is the reintroduction of wolves into the Greater Yellowstone Ecosystem beginning in 1995. At the same time, on the exclusionary side of the restoration coin, efforts to control or extirpate exotic species, including popular species such as deer, burros, and wild horses, have become a commonplace of restoration efforts, as has the public controversy these measures often arouse. Restorationists in the eastern part of the prairie triangle, where invasion of prairie by woody species has been nearly complete, have begun taking out trees, expanding prairies to create habitat for grassland birds such as Henslow's sparrow, sedge wren, and amphibians.[14] Others began worrying about earthworms when researchers realized that these had been extirpated by the most recent glaciation of the area and that all the species found in the northern parts of the prairie region are exotic—escapees or leftovers, their populations radiating out from popular fishing holes and establishing a silent, out-of-sight presence that can have marked effects on the vegetation and other components of an ecosystem.[15]

Literalizing the trope that ecocentric restoration means putting back not only the appealing elements but the poison ivy and rattlesnakes as well, the Eastern Massasauga Recovery Team, a consortium of agencies in Illinois, has begun a captive breeding program for massasauga rattlesnakes in order to ensure the future of declining populations at a number of sites.[16] Recently a consortium of soil scientists published a manifesto arguing for the importance of soils in restoration.[17] And, highlighting the ecocentric restorationist's commitment to authenticity, thinking about genetic provenance has become increasingly sophisticated as practitioners have added to their concern about genetic purity a complementary concern for the genetic diversity of populations used in reintroductions. This has been in part a response to conditions such as those in California, where, because of local variations in relief, exposure, and climate, genetic variation is thought to occur across short distances, but it also reflects growing concern that climate change might place a premium on genetic diversity in the foreseeable future.[18]

At the same time, as they have attended to the composition, structure, function, and dynamics of the ecosystems they attempt to restore, restorationists have also expanded their interests far beyond the tallgrass prairies and forests that provided models and inspiration for the earliest projects. In the past couple of decades practitioners have undertaken work on behalf of a wide range of ecosystems, from reefs and seagrass beds off coastlines to high-altitude systems and nearly every kind of terrestrial and aquatic system in between. Restoration efforts have been undertaken on sheer limestone cliff faces, in vernal pools in California, and even in caves.[19]

Laws and Regulations

Another step in the realization of ecocentric restoration has been passage of federal and state laws favoring various ecocentric versions of land management.

Broadly speaking, early legislation related to environmental management was utilitarian in its aims, designed to protect natural resources and promote soil conservation. The Pittman–Roberston Act of 1937, which supported management of habitat for game animals, is a classic example.[20]

Over the past quarter century, however, and roughly coincident with the emergence of ecocentric restoration as a conservation strategy, legislation and the regulations developed to implement it have embraced protection of nature for its own sake. These go beyond protection of natural areas to mandate restoration of features of damaged ecosystems even when, like the tree falling in the forest with no one to hear it, they have no economic value and their loss or impairment harms no one. One of the first laws to reflect this perception was the Wilderness Act of 1964, which described wilderness as places that retained their primitive character and influence and were to be managed to preserve this quality. The law did not promote restoration, but is important because it reiterates the cultural and ecological significance of historic landscapes, a concept that resonated with the idealism of Americans in the 1960s.[21] Even as Americans were taking advantage of this new law, legislation to protect endangered species was working its way through Congress. Ultimately the Endangered Species Act (1973) clearly promoted recovery of threatened and endangered species, but it is perhaps most important because it disallows economic reasons as justification for determining whether a species is to be listed as endangered or threatened. As J. Baird Callicott notes, the Endangered Species Act implicitly recognizes the intrinsic value of listed species and exempts them from purely instrumental, or economic, considerations.[22]

A watershed in this development was the passage in 1980 of the Comprehensive Environmental Response, Compensation and Liability Act (CERCLA), better known as the Superfund. Up until that time, Linda Burlington, who is senior counsel for damage assessment for the National Oceanic and Atmospheric Administration, notes that it was possible under existing legislation to sue for damage to the environment and even to appeal to the inherent value of elements that had been harmed, but that compensation was nevertheless pegged to economic harm to human stakeholders. CERCLA, however, established a per-barrel tax on oil to create a fund—the Superfund—that could be used to "make whole" damage to the environment, even when the person or corporation that had caused the damage was unknown or no longer existed. Under this law, Burlington notes, damages were still pegged to economic loss. But because these included loss of unspecified future values, this came close to granting legal status to nature for its own sake.[23]

Ecocentrism came out of the regulatory closet a few years later when the authors of the regulations implementing the Oil Pollution Act of 1990 pegged damages explicitly to the cost of restoration of the damaged system. Because this included damage to noneconomic elements of the ecosystem, it amounted to a legal mandate for ecocentric restoration. Burlington notes that this has met with general approval. "Even the corporations like it," she says. "They prefer it to just handing money over to someone. This way they actually see something for their money."[24]

Although the United States has no agency specifically responsible for restoration, it does have a program, the Natural Resources Damage Assessment and Restoration Program, which operates under Interior Department regulations emanating from CERCLA and the Clean Water Act. It requires "trustees"—federal and state agencies and Indian tribes—to administer compliance under CERCLA, the Clean Water Act, and the Oil Pollution Act. Once notified of a complaint, the agencies document damage in four categories—soil, water, air, and (most often) biota—and then propose correctives, which commonly entail some form of restoration.[25]

The process provides rich opportunities for the agencies to seek imaginative ways of getting maximum biological benefit from each dollar spent on remediation. As an example, Mike Hooper, a research biologist with the U.S. Geological Survey, notes that a plan to compensate for an oil spill that harmed habitat of the sooty shearwater in California in 1994 included a program to eradicate rats from the birds' breeding habitat in New Zealand, addressing what biologists suspected was a factor limiting the health of the population.[26]

The Business

Restoration as an avocation, an experiment, or a "Sabbath" exercise generates a certain kind of value all its own. But it doesn't meet the needs of a society on the other six days of the week. At some point, "getting real" entails taking on restoration projects as a job and a source of income for the practitioner.

Realizing this, and attracted by the notion that restoration (especially once it was backed up by legislation such as the Endangered Species Act and the National Environmental Policy Act) might provide an opportunity to do well while doing good, entrepreneurs began moving into the area, some jumping over the wall from academia to take advantage of this developing market niche.[27]

Andropogon Associates was among the first firms set up with this in mind in the 1970s and was soon followed by others. Ed Garbish left a faculty position at the University of Minnesota in 1970 and spent a year working for TNC's Center for Applied Research in Environmental Science before founding Environmental Concern, a private, nonprofit organization that pioneered wetland restoration. A year later, Michael Alder recognized the potential market for indigenous plant species and founded Native Plants, Inc., with headquarters in Salt Lake City. His venture paid off. The company was a financial success and maintains one of the largest collections of native plants and seeds in the world.[28] In the Midwest, Steven Apfelbaum launched Applied Ecological Services in 1978, specializing in prairie and wetland management. Others, already mentioned, including Robin Lewis, Andy Clewell, John T. Stanley, and George Gann, established for-profit or nonprofit enterprises during this period, from the mid- to late 1970s through the 1980s.

Companies have proliferated since, although exact numbers are not available. Recently, in his book *The Restoration Economy*, Storm Cunningham argued for the economic viability of restoration, which he describes as "the business and the spirit of the twenty-first century." Coming from a business background, Cunningham recognizes the economic potential of restoration as an economic opportunity. His book is optimistic, even promotional in tone, but as Barbara Bedford, echoing Eric Higgs's caution, warns, Cunningham's scenario could place ecological restoration at the mercy of modern investment capitalism. Like the Sierra Club's Michael Fischer, she worries that it might become a "subsidiary of destructive development."[29]

Steven Gatewood of the Society for Ecological Restoration noted in 2002 that the for-profit side of ecological restoration has clearly taken off. "People are doing restorations left and right," he added, though noting that the quality of the work varies from good to poor. Hundreds, perhaps thousands, of people are engaged in the practice.[30]

With so many people involved, the issue of professional standards has naturally emerged. John Munro, of Munro Ecological Services in Pennsylvania, has argued for adoption of professional standards. John Zentner, of Zentner and Zentner Land Planning and Restoration in Oakland, would go a step farther and open the work of consultants and practitioners to review. Standards might be simple, he suggests, but subject to review by others.[31] In 2000, the Society for Ecological Restoration published its *Guidelines for Developing and Managing Ecological Projects*, followed in 2005 by a revised edition. Such material notwithstanding, ecological restoration is not yet a formally recognized profession because it lacks a formal process for certification. But restorationists can take advantage of certification through related organizations, such as the Society of Wetland Scientists, the Ecological Society of America, and the Wildlife Society. In addition, the American Society of Landscape Architects participates with state governments to license practitioners.[32]

Chapter 10

Realization V: The Relationship

Beyond the bison, rattlesnakes, earthworms, and fire, what about the one species that has most conspicuously shaped and in many respects dominated the ecology of North America for the past dozen or so millennia: our own?

So far, we have considered the discovery and realization of the distinctive value of ecocentric restoration only in the ecological dimension, as a way of creating, re-creating, or maintaining historic ecosystems in order to ensure their well-being and perpetuation. We have said little about the discovery of the value of this work for the people involved as a distinctive way of engaging nature, learning about it, and establishing, shaping, recovering, and perhaps celebrating a relationship with it. Yet this is as important as its purely ecological value, if not more so.

Although early restoration efforts were focused—at least overtly—on environmental and ecological considerations, the emergence of restoration as a recognizable discipline and conservation strategy during the past quarter of a century or so has been accompanied by a broadening of perspective as restorationists with various backgrounds and interests began to incorporate a wider range of social, cultural, and ethical dimensions into their thinking and practice. *Restoration & Management Notes* had fostered discussion of these considerations from its creation in 1981, and they were well represented at Society for Ecological Restoration (SER) conferences from the beginning. In 1995 John Cairns Jr. wrote of "ecosocietal restoration," defining it as the process of reexamining society's relationship with natural systems so that environmental repairs and destruction might be balanced, and restoration might exceed degradation. Cairns also wrote that ecological restoration reflects social values, and philosopher Eric

Higgs articulated a similar idea in his essay "What Is Good Ecological Restoration?," arguing that restorationists should explore the value of their work in the dimensions of culture, ethics, morality, and aesthetics.[1]

That, of course, includes political considerations. "At its best," philosopher Andrew Light observes, "ecological restoration preserves the democratic ideal that public participation in a public activity increases the value of that activity." Local restoration projects that bring together human and natural communities illustrate this point. In the best examples, people participate as equals and create an egalitarian context for restoration. As Light puts it, "Our activity with nature is analogous to our activity with each other in a democratic society." He cautions, though, that restoration is not "inherently democratic"; "rather, it has an inherent democratic potential that might be lost or preserved in any act of preservation."[2]

Obviously, this "activity with nature" defines a relationship with the environment and provides opportunities for the creation of transcendent values, such as community and meaning, that grow out of any relationship. And because the condition and fate of an ecosystem ultimately depend on how the people who inhabit it behave—that is, on how they see the world and on what they know and think and care about—this is crucial. As we noted in chapter 1, humans have been apprehensive about the degradation and "running down" of nature for as long as we have any record, but they have understood this "running down" as more internal and subjective than external and objective—as less about the actual deterioration of ecological systems than about the decay of the knowledge and values that humans create and use to make sense of the world and negotiate their relationship with it.

This being the case, "world renewal" in traditional societies does not mean land management projects that change the land but rituals such as the Intichiuma of the Australian Aborigines or the Sun Dance of the Plains Indians that shape, renew, and transmit the intellectual, emotional, and spiritual software that defines the relationship between the land and the people who inhabit it.

As it happens, the discovery and realization of the value of restoration as a context for this kind of work—affecting the mind and the soul as well as the land—have gone on hand in hand with the discovery of its ecological value. In this chapter we will provide an overview of this development in several categories: the discovery of the value of restoration as a way of experiencing the natural landscape; as a context for learning and a technique for basic ecological research; as a way of building community; and finally and most broadly, as a context for negotiating and articulating the terms of the relationship between the human community and its environment.

In this last category especially, we will be closing a circle, considering how practitioners have learned to exploit the experience of ecocentric restoration—a technology just about as old as the airplane—as a way of carrying out one of humankind's most ancient tasks.

Community-Based Restoration

Any action, from cooking dinner to restoring a prairie, not only creates a product but also creates value—or disvalue—in other dimensions as well. We may call these process, experience, and performance, and the development of ecocentric restoration has entailed the discovery of value in all these dimensions. Indeed, a major reason for Aldo Leopold's influence on conservation over the past three-quarters of a century has been that, in addition to his scientific publications, he regularly abandoned the stilted, passive-voice rhetoric of the scientific paper to write of his own experience as a conservationist and even restorationist, as he did most notably in *A Sand County Almanac*. Such accounts of restoration have been rare, however. Although a handful of practitioners have published personal accounts of the experience of restoration in recent years,[3] the clearest expression of the value of this work for the people involved is not yet in print but on the ground, in the emergence of the school programs and citizen-based restoration projects that have proliferated during the past few decades. As this has happened, it has constituted a kind of revolution in the organizations and communities involved. Because it opens up a way to connect large numbers of people with old ecosystems, linking ecology with sociology and history, this was precisely the revolution needed to turn restoration from what Higgs calls "technological restoration" into an occasion for negotiating a community's relationship with its environment. The result has been what U.S. Forest Service (USFS) researcher Herbert Schroeder calls a "volunteer ecological restoration movement" involving thousands of people who work to restore endangered native ecosystems in their neighborhoods or in parks, preserves, and other public lands.[4]

One of those who played a key role in this development was Steve Packard, who, beginning in the mid-1970s, undertook in the Chicago area one of the earliest and certainly one of the most influential community-based restoration programs in the country. Packard was by no means the first to welcome or recruit volunteers to help with restoration efforts.

There is a long history of volunteer involvement in conservation efforts, and as we have seen, Paul Shepard had recruited volunteers to help with the restoration effort at Green Oaks twenty years earlier. Bob Betz relied heavily on volunteers to help with his project at Fermilab, and the early restoration efforts in the Pacific Northwest were driven entirely by citizens taking to the woods and streams as individuals and as members of informal groups or voluntary organizations.

Packard approached the restoration business from an unusual angle, however. He had developed formidable skills in grassroots organizing as a young man working in the peace, civil rights, and social justice movements, and in many ways his work in restoration has been a continuation of that work. From the beginning he made community participation a high—even top—priority in this effort, with the self-conscious aim of developing nothing less than a culture of restoration. The result has been a program that has gained wide attention and has served as a model for similar projects in other parts of the country.

Living in Chicago in the early 1970s and looking for ways to redirect his energies as the era of war protests came to an end, Packard turned his attention to the environment. Looking back, he recalls a moment of epiphany when, his head full of the celebration of the old prairies he had read in an essay by Bob Betz,[5] he studied the local flora in his spare time and eventually came across one of the remnants of prairie that lie scattered along the North Branch of the Chicago River. What struck him was not only that this relic of an all-but-vanished landscape still existed but that it was clearly in sorry shape ecologically, rapidly being invaded by brush and obviously in need of help. Inspired, he started tinkering with restoration himself and was soon showing up at meetings of the local unit of the Sierra Club and other organizations to recruit volunteers for weekend sessions cutting brush, gathering seed, and, eventually, managing burns on various prairie remnants scattered through Chicago and its suburbs. Before long, he landed a job as director of public information for the Illinois Nature Preserves Commission. Schooled as he was in organizing, he approached this work much as he had his earlier work in the "movement," placing a high priority on the people side of the nature–culture nexus.[6]

"I argued that if our aim was to preserve these places, we needed a constituency for them," he recalls. "And I thought that getting people involved in helping to restore them would be a powerful way to develop such a constituency." Attracting recruits for the cause of restoration in Chicago's extensive forest preserves, Packard helped set up a Volunteer Stewardship Network that by the early 1990s had more than four thousand members, dozens of local units with regular newsletters, and projects at some sixty sites totaling some 2,000 hectares in the forest preserves and other set-aside areas. Packard left the Nature Preserves Commission to work for The Nature Conservancy (TNC) in 1983, and over the next decade he participated in what regional director Russ Van Herrick later called a "revolution" within the conservancy, based in large part on its discovery of restoration and the role volunteers can play in the restoration effort. As Packard recalls, Greg Low, who was TNC director at the time, endorsed the work Packard had been spearheading in Chicago. Low approved trying it out on an experimental basis at a number of TNC sites, and this work, increasingly backed up by newly emerging organizations such as the Natural Areas Association and SER, soon began making an impression on the organization.[7]

As in other organizations, this met with some resistance. Michael Reuter, who is currently senior director for conservation strategies for TNC's Central United States unit and director of its Great Rivers Partnership, recalls that this was "a period of pretty intense struggle in the Conservancy. There was concern about mission drift, and a lot of reflection and intense discussion about whether restoration could really achieve aims in line with the Conservancy's mission, or whether maybe we should just focus on intact systems." But a lot had happened, both on the ground and in managers' minds in the decade or so since Bob Jenkins had walked into his office at TNC to find an organization incapable even of monitoring its preserves, much less restoring them. And Reuter seconds Packard's characterization of what had happened in TNC by the end of the 1990s as, if not a revolution, then certainly a development that has "taken us into a whole new way of thinking, not abandoning what we had been doing, but building on it."[8]

Specifically, the notion of restoration as an activity that could benefit rather than suffer from involvement by large numbers of citizens opened TNC up to the realization that volunteer programs might make it possible not only to maintain preserves but also to expand them, reversing the decline in the quality and extent of preserves that conservationists had pretty much regarded as inevitable at least since Aldo Leopold had written that "wilderness is a resource which can shrink but not grow."[9] At the same time, it offered something environmentalists had been seeking for generations: a basis for a relationship between contemporary people and natural ecosystems that is both active and positive in its effects.

As far as ecology alone was concerned, it meant that TNC could start "looking beyond its fences," as Reuter says, acquiring degraded as well as pristine lands, with the intention of expanding and upgrading existing preserves. The inclusion of the human element, represented by the kind of work Packard was doing in Chicago, made it clear that the human inhabitants of an area, far from being a threat to this mission, could actually play a key role in carrying it out. And because this not only would help with the work but would actually change the sign of the relationship between humans and the natural landscape from negative to positive, it really did amount to a revolution, one that was taking place not only in TNC but in organizations, agencies, and communities all over the country.

Liberated in this way from a narrow, hands-off preservationism, TNC began buying larger tracts of land, often including areas in need of intensive restoration, taking on projects on an expanding scale that has paralleled the logarithmic expansion of prairie restoration projects during the past half century. Reuter cites as notable examples the Nachusa Grasslands project in western Illinois, which topped 3,000 hectares in 2004, the 2,800-hectare Kankakee Sands Grassland in Indiana, a 15,600-hectare tallgrass prairie project in Oklahoma, and similarly scaled floodplain projects in Illinois and Louisiana, all begun in the 1980s and 1990s.

In the meantime, the volunteer-oriented, citizen-based approach to restoration pioneered by Packard and the North Branch volunteers has flourished on many TNC preserves. At Nachusa Grasslands, for example, project director Bill Kleiman has developed a program heavily dependent on volunteers, who take responsibility for tracts ranging from a dozen to as much as 50 hectares. Although some programs depend on large numbers of volunteers working around the margins of the work week, Kleiman describes his program as "an inch wide and a mile deep," with a handful of volunteers who regularly spend several days a week on their sites, working out of an old farmhouse at the preserve in a kind of monastic seclusion. He notes that one couple recently took advantage of the buyer's market in real estate to buy a small house near the preserve so they can spend more time there working on their unit, a gas-saving variation on the tradition of driving up to Wisconsin to spend time at a vacation home on a lake. He also points out that the Nachusa volunteers are working on land owned by TNC and held in public trust, an act of social and biological altruism that fosters a high level of community—and commitment—among the participants.

Community-based, volunteer-oriented restoration programs have taken shape in other parts of the country in the past couple of decades. In the San Francisco Bay Area, early initiatives such as those at Strawberry and Wildcat creeks have proliferated, with projects under way in a growing number of neighborhoods and related programs of study and internships at Merritt Community College in Oakland, UC–Berkeley, City College in San Francisco, and the Presidio and other units of the Golden Gate National Recreational Area, a complex of open spaces maintained by the National Park Service. According to Pete Holloran, who was an intern with the Presidio's stewardship program in its early years, projects such as these are generally underrepresented in the published literature but have served as models for similar projects in a wide range of situations. As an example, he points to the "green conversion" of military bases closed since the end of the Cold War.[10] Certainly agency skepticism about restoration and participation by volunteers has declined dramatically in recent years, in some cases evaporating entirely and being replaced by enthusiasm. When Ed Self, now director of Wildlands Restoration Volunteers in Boulder, surveyed USFS staff for their views on the role of volunteers in restoration work in 2000, he got a strongly positive response, with respondents generally reporting that they regard restoration as important on USFS lands and that they believe that the benefits to the volunteers are an important consideration in designing their volunteer programs.[11]

Motives

As programs such as these have evolved in recent years to the point that they have become a conspicuous feature of conservation, they have prompted a good deal of reflection and research into what motivates participants and the value of these programs, not only for the ecosystems being restored but also for the people involved.[12] Ed Self, for example, drawing on his experience with the Wildlands Restoration Volunteers, notes that volunteers often express satisfaction with their experience in restoration and in building a "caring community of land stewards," transferring "the healing metaphors of restoration beyond the task at hand to other parts of their lives."[13] Peter Leigh, who has embraced volunteer-based restoration in connection with his work in the National Oceanic and Atmospheric Administration's Habitat Conservation Office, notes that it offers participants a chance to meet "their obligations to future generations, other species, and entire ecosystems by redeeming both nature and humanity by restoring places of beauty and ecological importance." He sees community-based restoration as a form of environmental therapy that allows citizens, including those living in cities, to "participate again with the rest of the living kingdom in our effort to recapture the 'garden.'" And John Thelen Steele, reflecting on his experience with community-based projects in the San Francisco Bay Area, notes that reconnecting with nature in this way is a way to heal our anxious and isolated psyches.[14]

Responding to these observations, researchers have begun to explore the experience of volunteers involved in restoration efforts more systematically. Herbert Schroeder, an environmental psychologist with the USFS's North Central Research Station in Evanston, Illinois, examined newsletters of local groups involved in restoration in the early 1990s and identified altruistic aims, along with a sense of the vulnerability of nature and the urgency of responding to environmental degradation, as consistent themes in these written expressions of the experience of volunteer restorationists.[15] Other researchers have identified a general interest in nature, the opportunity to learn about local ecology and natural history, the chance to get away from everyday concerns, and the chance to socialize as important motives for volunteers. And, addressing the important question of whether these experiences are just a form of joint effort by the environmental choir, researchers have also found evidence that the experience does lead some participants to adopt more environmentally friendly behaviors.[16]

Others have considered this development from a political perspective. Bill Jordan has pointed out that, as a kind of environmental "junkpicking," restoration is sociologically inclusive. For one thing, it offers opportunities to "do nature" near where most people live. For another, because it creates natural value rather than consuming it, it makes room for and indeed welcomes large numbers of participants, in contrast with consumptive—and therefore exclusive—ways of enjoying nature ranging from hunting and fishing to birding and backpacking.[17]

Of course, if participants are the actors in restoration, understood as a political act or performance, they also have a constituency—or audience—made up of the general public. Because much of the value—or disvalue—of restoration depends on how it plays for this larger public, this is an important consideration, and investigators have looked into it as well. When David Ostergren and his colleagues surveyed public views about forest restoration in the Flagstaff area, they found that attitudes varied widely. Although most of 663 respondents agreed with the proposition that forest fuel loads had to be reduced (understandable in a town surrounded by ponderosa pine forests), most also acknowledged that ecological restoration involves active management and includes more than reducing the risk of fire. As might be expected, responses indicated a mix of altruistic and self-interested attitudes. There was strong support for the idea of recovering rare and endangered species, and a large majority (87.1 percent) agreed that restoration meant restoring "working ecosystems"—certainly not incompatible objectives in this case. But the survey revealed a striking divergence of opinion about the practice of removing large trees—or most trees, for that matter, large or small.

Similarly, Peter Friederici, in his study of the "Flagstaff model" of forest restoration, identified a wide array of perceptions. Writing in 2003, five years after this project began, he concluded that restoration does not have a single meaning in Flagstaff but rather multiple meanings, defined by stakeholders' differing perceptions and demands on the land.[18]

Indeed, in an earlier study in the Chicago area, researchers Susan Barro and Alan Bright documented a striking disparity between public attitudes toward the *idea* of restoration and the *act* of restoration as it actually played out in places such as the area's forest preserves. They suggest that this results from the difference between the denotation of the word *restoration*, which is value-neutral, indicating only a return to a *previous* condition, and its strongly positive connotation, which implies a return to a *better* condition. When it turns out that restoration in the Chicago area commonly entails cutting down trees, reintroducing fire, and possibly even shooting deer in order to bring back ecological systems that most Illinoisans are unfamiliar with and many find unappealing, many withdraw their approval.[19] This finding is directly relevant to the distinction we are making between ecocentric restoration, which may entail setting aside specifically human interests, and meliorative restoration, which is aimed at making a "degraded" landscape or ecosystem "better," leaving open the questions, "Better in what way? And for whom?"

Backlash

These questions and the often divergent ways people answer them obviously have important political implications. Although community-based restoration may hold a certain democratic potential, as Andrew Light suggests, and may also provide a way to address our ecopathology, as Peter Leigh indicates, disagreements about aims, methods, procedures, and even the definition of restoration have sometimes hindered the best efforts, as indicated by two well-known incidents, one in Chicago, the other in Missoula.

In the spring of 1996, the vibrant, well-established restoration community that had developed in the Chicago area during the previous two decades was rocked by a wave of public objections. Later described by a member of that community as "a public relations disaster,"[20] this was a backlash by citizens who objected to the restoration effort and argued that they had been left out of decisions about the management of publicly owned lands. Sparked by citizens who objected to restoration efforts in wooded areas in their neighborhoods, which often entailed cutting trees, reducing shade and opening up unwanted sight lines, and fanned into flame by *Chicago Sun Times* columnist Raymond Coffey, the protest quickly developed into a firestorm that led at one point to a moratorium on restoration efforts in two counties that lasted several months and set back restoration efforts in some areas for years.[21]

Analyzed with great care over the next few years by observers on both sides of the argument, and by researchers aiming for a measure of impartiality, this "Chicago controversy" proved to be a coming of age for Chicago's nascent restoration culture. The controversy brought into the public forum questions and contradictions inherent in the notion of restoration, including the question of the proper role of humans in nature, the value of attempting to return a piece of land to a previous condition, the ethics of favoring one suite of species—the "natives"—over another, and, underlying the whole debate, questions of altruism and self-interest that ultimately distinguish ecocentric restoration from other forms of land management. Eventually settling down after the stay on restoration was relaxed, it nevertheless left restorationists in the region badly shaken, causing some of them to question their commitment and the validity of the whole idea behind their effort.[22]

Such controversies are hardly unique. In her overview of the Chicago blowup, Debra Shore describes similar controversies over the removal of Australian pine from coastal ecosystems in Florida, measures to restore the historic structure of woodlands in California, and objections by pigeon fanciers to the return of peregrine falcons in New York City. Nor are these controversies likely to go away because, as sociologist Reid Helford points out, restoration, which involves changing people's environment, is inherently political.[23] After all, people are God's creatures, too. Like other creatures, they have conflicting interests that complicate relationships between them, and the management of relations within the human community is as important as purely ecological concerns in planning and carrying out restoration projects.

Value diversity of this kind naturally complicates the process of planning and implementing restoration efforts, and indeed a recent study of restoration efforts on behalf of ponderosa pine forests in the Southwest shows that "the pace of implementation has often been alarmingly slow." The authors cite three barriers: insufficient funding, social conflict over the meaning of restoration, and the difficulty of defining the nonmarket benefits of the restoration. The region is culturally complex and has only recently made a transition from a resource extraction economy to one driven by tourism and recreation. Furthermore, a large portion of land in the region is publicly owned and has a history of escalating wildfires. All these factors have contributed to social discord regarding restoration activities and probably have stalled community-based projects. One of the most intractable debates in northern Arizona resulted from the different ways in which scientists, environmentalists, and some community officials define restoration. These led to more basic disagreements, because as the authors of this study note, "By its very nature, ecological restoration touches on fundamental questions about the relationships between people and their environment, and is therefore fertile ground for conflict."[24]

But also, at times, resolution. In 1996, the University of Montana undertook restoration of weed-infested Palouse prairie on Mt. Sentinel, which rises on the eastern edge of the campus. Restoring this rare example of a low-altitude grassland in the Rockies meant removal of exotic vegetation, and a plan to use herbicides to accomplish this provoked public objections, leading the university to reject the plan.[25] The following year it approved a plan that called for minimal use of herbicides and emphasized public participation in every stage of the effort.[26] "We stopped talking about 'killing weeds' and began talking about 'restoring native grasslands,'" the authors of an account of the effort wrote.[27] A sign of the restoration plan's success came in June 2000, when county voters passed a property tax increase to fund control of noxious weeds on the site.[28]

Both conflicts and successes of this kind dramatize the democratic potential of restoration. This is what democracy is about: not agreeing, but debating conflicting values and views. It is in this spirit that observers such as Reid Helford emphasize the importance of the social and political aspects of restoration efforts. Resource analyst Kimberly Botsworth Phalen has explored the value of the Reasonable Person Model, a theory of human behavior, motivation, and cognition, both for analyzing disputes such as the one in Chicago and for avoiding such blowups in the future. She contrasts this controversy with the successful effort at Montrose Point in Chicago's Lincoln Park, which occurred in the same area at the same time. Here, the key was effective two-way communication that produced a useful cognitive map for both managers and stakeholders.[29]

This brings up an interesting point with respect to the distinction we are making between ecocentric and meliorative land management. Meliorative land management comes in as many forms as there are ideas of improvement and self-interest, and a healthy polity will provide a context in which these interests can be debated in a productive way. The outcome of such a debate is usually some kind of compromise: new plantings along the lakeshore, for example, but with more room for a soccer field than proponents of the plantings had in mind. Ecocentric restoration is obviously just one of many such land management models a community may adopt, and it, too, allows for compromise: Emphasize native plants, but arrange them in attractive ways and omit the ragweed and poison ivy. In the strictest sense, however, ecocentric restoration precludes such compromises and demands that we not creatively imitate but humbly copy the model system.

This being the case, ecocentric restoration is best regarded in a democratic context as one of many options for land management, not necessarily better than any other, but offering its own kind of benefits. Having debated these options, a polity—a city, a watershed cooperative, or a neighborhood association—may decide to take on this exercise in deference to nature, or not. Having investigated the debates surrounding restoration in the Chicago area in some detail, Paul Gobster points out that conflicting ideas about land use, including those based on divergent ideas of nature, can often be accommodated by nesting iconic features cherished by different interest groups together in a single project.[30] But once a community has dedicated a piece of land—perhaps a bit nested in a conventionally landscaped park—to ecocentric restoration and has decided on the model system, all that remains to be debated is economics and logistics, and the project has to be guided by experts, including, of course, those with intimate place-based knowledge, as well as professional historians and ecologists. In this sense, ecocentric restoration is a top-down and, if you like, elitist idea. It is not one that emerges from direct experience of the land, however intimate, unless that experience is informed by knowledge of the history and ecology of the site. This is presumably what one of Reid Helford's informants meant by insisting, "You can't have compromise on what nature is in northeastern Illinois. I think that's really the bottom line. . . . It is what it is. That's not something you can change by talking about it."[31]

This clearly implies a certainty, reminiscent of Clement's idea of an ecology outside of history, about what nature is that neither science nor history will deliver. But it is this insistence on the old ecosystem as a model, and the Sabbath-like act of deference to the best available description of this model it implies, that gives ecocentric restoration its distinctive psychological, moral, and ecological value. However, this value finds little resonance in an individualistic culture that is not only skeptical about authority but resistant to the very idea of obedience—even, perhaps, to nature. Ecocentric restoration is different from the self-interested games humans spend much of their time playing with the rest of nature, and the distinction is important. Indeed, it may well be that conflicts (as opposed to debates) over proposals to restore a piece of publicly owned land occur in large part because those proposing restoration often overlook or downplay this distinction, implying that what is good for the model ecosystem will also be good for the public in some fundamental, generalized sense—better, in fact, than something else the public might actually want.[32] This, of course, is always a debatable proposition.

Restoration Ecology

The notion that one way to learn about something or test one's ideas about it is to take it apart and then attempt to put it back together is a fairly obvious one. This being the case, it was only a matter of time before those who had taken on the task of restoration began to become aware of and to draw attention to the value of this effort as a way of raising questions and testing the ideas about the composition, structure, dynamics, and function of the model ecosystem. At the UW–Madison Arboretum, ecologists John Curtis and Max Partch, reflecting on their early experiments on the use of fire in restoring the arboretum's prairies, had noted in 1948 that "much information of value concerning the dynamics of formation boundaries can be obtained in the course of such establishment."[33] In fact, synthetic approaches to ecological research have a long history, and the arboretum's restoration efforts have created opportunities for a good deal of research, including research that could not have been done in pristine systems. It was in reflecting on this experience and drawing on writings of a number of biologists, including entomologist E. O. Wilson and British ecologist Anthony Bradshaw, that in 1977 Bill Jordan and his colleague Keith Wendt formalized the idea of restoration as a technique for basic ecological research and gave it a name: *restoration ecology*.[34] Following up on this a few years later, Jordan worked with ecologists Michael Gilpin and John Aber to organize a symposium to explore this idea and its implications for both restoration and ecology.[35]

As we have noted, the idea of learning about ecological systems by assembling them was not new. Researchers have been creating ecological systems in various ways, both in the field and in the laboratory, for a long time.[36] And in 1983 ecologist Anthony Bradshaw described restoration itself as an "acid test" of ecological understanding.[37] But since the value of restoration as a way of raising questions and testing ideas about ecosystems had not yet been explored in a systematic way, this process of getting together to discuss this idea and giving it a name was a significant step in the realization of its value. One aim of this initiative was to valorize restoration intellectually by proposing that it is not—or is not only—a form of applied ecology and that the relationship between basic and applied research in ecology could be dialectical rather than hierarchical, and this seems to have had some effect. Looking back, ecologist Joy Zedler, whose research on wetlands has entailed various forms of restoration ecology, says she believes that it helped rescue restoration from the low-class status it had as a form of applied ecology, giving it a place in the ecologist's repertory of techniques and concerns and contributing to a turnaround in attitudes that has occurred in the years since. She notes that it sparked discussions that led to the development of courses in restoration at San Diego State University, where she was teaching at the time, and points to a parallel

Two Cultures?

The question of relations between research and practice is a bit like the question of the weather. People talk about it a lot and are often unhappy about it. But is there really a problem? And, if so, is there anything much that can be done about it?

Curious about this, Kim Frye, who recently completed a master's degree in urban ecology in the Environmental Sciences program at DePaul University, hunted up some thoughtful people with experience on both sides of the aisle and asked them about their experience in the borderland between theory and practice. The responses varied widely, suggesting that the relationship between researchers and practitioners reflects the diverse contexts in which restoration projects are carried out.

Don Falk, of the University of Arizona's School of Natural Resources, noted that in his region, although researchers and managers may argue about the details, they agree on what he regards as the key idea: that the reintroduction of fire is necessary for the restoration and maintenance of many of the region's historic ecosystems.

Others emphasized what they see as a disjunct between researchers and practitioners, some attributing this to circumstances, others to a fundamental difference in the aims of the two vocations.

Two Cultures?
Continued

Joy Zedler, for example, noted that she undertook her pioneering research on tidal wetlands in the San Diego area to answer practical questions raised by attempts to recreate wetland habitat as part of the mitigation process but that these questions were also of great ecological interest. Yet despite this long experience of research that had both practical and theoretical value, she noted "a huge prejudice against applied science in the universities." "I was lucky," she told Frye, "to be at a university [San Diego State] where that bias wasn't present. I didn't encounter it until I moved back here" (to take, a bit ironically, the position of Aldo Leopold Professor of Restoration Ecology at the UW–Madison).

This bias exists on both sides of the culture gap. Dennis Nyberg, director of the James Woodward Prairie at the University of Illinois–Chicago, said that although he sees science as a way of gaining useful knowledge, in his experience in the Chicago area restoration practitioners are generally not much interested in that but mostly "learn a few simple things and then stick with them," their efforts reflecting more "wishful thinking" than a serious interest in the efficiency or long-term effectiveness of their work. At the same time, despite having devoted much of his energy to restoration-related research, he noted that restoration poses many challenges for the researcher. As he told Frye, "I don't see the techniques of the ecologists as being very useful in general."

Whereas Nyberg attributes this problem to the technical, logistical, and organizational factors that complicate the work of the restoration ecologist (notably the complexity of the task and the time it takes to get useful results), Tom Simpson, restoration ecologist for the McHenry County (Illinois) Conservation District, suggested that it reflects a fundamental difference between a culture committed to learning about phenomena in an abstract, universalizing way and a culture concerned about restoring actual ecosystems. The result, he has suggested elsewhere, is what he believes is a serious misunderstanding: "We have defined restoration ecology as a subdivision of ecology rather than a subdivision of restoration," he says, "and, not surprisingly, it serves ecology and ecologists more than restorationists."[a]

Simpson argues that recognizing that research and practice are different in fundamental ways is the key to forging a productive relationship between them. DePaul University ecologist Liam Heneghan points

---

Two Cultures?
Continued

out that that is pretty much what Chicago Wilderness planners had in mind when several years ago they divided the organization's Science and Land Management Team into separate bodies, one dealing with research, the other with management. The idea was to clear the way for "authentic science," ideally in a useful relationship with practice. This is still "mildly controversial," Heneghan notes, and it's too early to assess the results.

However, managers have welcomed a recent initiative to undertake an integrated program of social science and natural science research on a number of sites in the Chicago area, and Heneghan notes that in his experience amateurs working as volunteers are more likely than professional managers to resist collaboration with researchers.

He recalls a project that fell through a few years ago when the site manager decided she was uncomfortable with a plan to incorporate an experimental design into ongoing restoration efforts on her site. "She said it made her feel they were experimenting with thalidomide on babies," he said.

---

a. Thomas B. Simpson, "A Science of Land Individuals," *Ecological Restoration* 27, no. 2 (2009): 115–21. The item quoted is on p. 119.

---

development in the Ecological Society of America, noting that over the past quarter of a century the society has broadened its scope far beyond basic research and now places emphasis on challenges such as species conservation, reintroductions, and the ecology of invasive species, as well as restoration.[38]

Restoration at School

By the early 1990s, proponents of ecological restoration were acknowledging—perhaps still a bit tentatively—that their fledgling discipline had arrived. Around this time, Bill Jordan recalls Bob Betz opining that ecological restoration was going to be a "big thing." "I'll bet one of these days they will be teaching courses about this at universities," he added with characteristic enthusiasm. The statement was nothing if not prescient, as institutions with research and teaching strengths in ecology, agriculture, landscape architecture, forestry, rangeland management, and related disciplines soon developed programs in the new field. Some have established restoration institutes within existing colleges or departments, their programs typically reflecting the ecology and environmental challenges facing their region.

A prominent example, not surprisingly, is the UW–Madison, where a concentration in the restoration and management of native vegetation launched by landscape architecture professor Darrell Morrison in the 1970s was among the earliest programs of its kind in the United States. More recently, an endowed professorship, named in honor of Aldo Leopold and first proposed by Bill Jordan in 1978 as a way of institutionalizing the idea of restoration ecology, now serves as a centerpiece for restoration-related programs in a number of departments.[39]

But Wisconsin is by no means alone. In 1973, Luna Leopold introduced "Hydrology for Planners" at UC–Berkeley, a foundation course still offered for students pursuing restoration studies.[40] Elsewhere around the country and in Canada, restoration studies have proliferated and are part of university curricula at some thirty institutions. A few—including Wisconsin, UC–Davis, Ohio State, Washington University, and SUNY Buffalo—offer a doctorate in restoration ecology or a related field. Master's and bachelor's programs are more widespread, although they probably do not exceed a dozen. More common are ecological restoration courses, institutes, and certificate programs, numbering at least several dozen.[41] North of the border, the Restoration of Natural Systems program at the University of Victoria in British Columbia, created in 1996, is among the earliest restoration education programs. It now offers more than a dozen regular courses to more than forty students each year in the diploma program, and Eric Higgs, who is on the faculty there, characterizes it as one of the top programs on the continent.

Surveying restoration programs online reveals little about the inspiration behind their development or the philosophical context that guides them, however. Most describe themselves in terms that suggest a generalized idea of restoration, focusing on the restoration of natural resources but also reflecting an interest in the restoration of biodiversity and whole ecosystems. Clemson University's Restoration Institute was established in 2004 "to drive economic growth by creating, developing and fostering restoration industries and environmentally sustainable technologies in South Carolina," clearly emphasizing economic benefits. At the same time both the institute and North Carolina State University describe their approach to stream restoration as "holistic," a word that suggests a commitment to restoration of "all the parts,"[42] and a similar, integrative philosophy seems implicit in the way other programs describe themselves in catalogues and on websites. Programs at UC–Davis reflect both utilitarian and ecocentric rationales for restoration, and a broadly interdisciplinary curriculum ensures a holistic context for study in this area. Writer Peter Friederici of Northern Arizona University's Ecological Restoration Institute, notes that "restoration must benefit and sustain human communities" if it is to remain a part of the social landscape, while Peter Fulé, also associated with the Ecological Restoration Institute, acknowledges that although restoration projects are typically motivated by some form of self-interest, ecocentric ideas are often involved.[43] At the Colorado Forest Restoration Institute, director Dan Binkley comments that, as far as ecocentric and self-centric objectives are concerned, he finds that most people "like to resonate with both."[44]

Restoration programs for K–12 students are hard to quantify, but many now engage youngsters in activities that are at least consistent with an ecocentric approach to restoration. The University of Wisconsin's college-level program is complemented by the UW–Madison Arboretum's K–12 education outreach program, Earth Partnership for Schools. This program, an outgrowth of a project launched by arboretum ranger Brock Woods and Bill Jordan in the early 1990s as an extension of the idea of restoration ecology beyond formal research into education, provides teachers with hands-on instruction in how to incorporate ecological restoration into a curriculum. In California, the "Kids in Nature" program at the Cheadle Center for Biodiversity and Ecological Restoration at UC–Santa Barbara includes the planting of native plant gardens on school grounds. Other noteworthy programs include the collaborative effort called "Restoring Gary's Urban-Industrial Forests" in Indiana, programs created by the Marianist Environmental Education Center at Dayton, and Bowling Green State University's Center for Environmental Programs, a joint effort with Toledo's Metro Parks.

Performance

If it has taken some time for restorationists, constrained by the prevailing myth of objectivity, to discover the value of restoration as an experience and to begin exploiting it in a systematic way, it should come as no surprise that they have been even slower to discover or acknowledge its value as an expressive act and to develop it explicitly as a performing art and context for the investigation, expression, and creation of meaning—the fullest development of its value as a form of play. Far from trivializing restoration, this is the dimension of experience in which its value is most fully realized. It is also the dimension in which the distinction between ecocentric and meliorative forms of land management is most salient, distinctions being the essential foundation for the creation of meaning.[45]

Looked at from this perspective, restoration has a lot going for it. In the first place, unlike the nonact of preservation, it is active—that is, an *act* in the literal sense, which is prerequisite to its development as an act of imagination. It also entails dramatic, emotionally charged actions, such as setting fires, and both nurturing and killing plants and animals, which implicate the restorationist in natural processes in an emotionally compelling way. To the extent that it is impractical, it is properly relegated to the dimension of play rather than work; no one but a fool works at something she understands to be impossible, but everyone plays at such adventures. Indeed, to the extent that it is impossible, it is consistent with what theologian Catherine Pickstock has called "the impossibility of liturgy," the acknowledgment that what liturgy (that is, public ritual) attempts—in the case of restoration, the reversal of time—may indeed be impossible.[46]

Besides this, unlike activities such as birding or botanizing, restoration provides rich opportunities for group effort and so provides a context not only for personal ritual but also for shared, public ritual. Indeed, restorationists working in groups often develop various performative elements such as gathering at a regular time each week, often on Sunday, and developing protocols such as moments of group reflection (including a certain food item in the windup meal, for example) that add value to the experience. (Those involved often take these ritual elements very seriously. A restorationist in the Chicago area told Bill Jordan that a visitor who joined her group on one occasion to pick up ideas for forming a group of her own called back after the visit to ask what kind of bagels they used.) But, as in the case of restoration ecology or restoration itself, the value of such activities depends in large part on how they are framed and articulated and the extent to which they are consciously experienced *as* performance or ritual, and social conditions in the United States offer formidable barriers to the systematic realization of the value of an activity such as restoration in the dimension of performance and ritual.

Of these, perhaps the most obvious is the diversity of American society, which calls for the development of a serviceable repertory of behaviors and expectations that can serve as a kind of cultural lingua franca, defining the protocols of social behavior out on the street among people with a wide range of cultural backgrounds. These are necessarily low-key rituals and go only so far in carrying out the intimate and psychologically fraught work of negotiating the relationships that are the source of identity, meaning, and community—work that for most Americans is carried out, if at all, in the context of elective enclaves such as service clubs, gangs, or communities of faith.

This being the case, unless a group of volunteers coming together to work on a restoration project is composed of members of an organization of this kind, it is likely to be difficult to establish the understanding—or performative contract—that is the basis for a performative community prepared to go beyond the low-key rituals of the street to a higher level of performative engagement and intensity. Thus California restorationist Pete Holloran points out that, in his experience, leaders have actually backed away from including performative elements in volunteer-based restoration projects over the years as they have realized that new recruits were often put off by rituals developed by a culturally homogeneous founding group. "That was keeping newcomers out," he comments, "and that wasn't doing us any good. What we want is a big tent."[47]

Besides the simple fact of cultural diversity, the prevalent culture is deeply imbued with a skepticism about the efficacy—and even the legitimacy—of performance that came over quite literally on the *Mayflower* with the Puritans, who were to a large extent defined by their rejection of the rituals of the pre-Reformation church. The resulting conviction that the "restoration of first times" that we mentioned in chapter 2 would be carried out literally rather than through the renewal and reconstruction of ideas, values, and emotional structures survives and lends environmentalism its characteristically robust commitment to social action and reform: Save the Whales! Plant a Tree! Restore a Prairie! Recruit Volunteers! Write Your Congressman! But this literalism entails a weakening of a culture's grasp of the very technologies of the imagination that are a defining feature of just those premodern, mythopoeic societies environmentalists have often held up as examples of a sustainable, gracious relationship with the environment.

This predisposition, or ordering of priorities, against play and in favor of work is evident not only in a failure to fully realize the value of restoration as an expressive act but also in the response of scholars and "practical" people alike to the very idea of restoration as a performing art. Bill Jordan recalls a conference in Ohio some years ago in which he ventured the notion that upgrading the performative aspects of restoration projects might add value to them, both for the participants and for their audience. What he had in mind was basically introducing protocols in the handling of tools and perhaps finding and taking advantage of the gestures and rhythms of activities such as brush cutting and seed collecting in order to enhance the experience of the work. But the response of some was, "Hey, this jerk wants us to put on clown suits when we burn a prairie."[48]

In a friendlier and more nuanced response, Eric Higgs, who has been contributing reflections to the conversation about restoration for more than two decades, has explored the idea of restoration as a "focal practice"—that is, things we do mindfully and for their own sake, such as holding a celebratory dinner or reading to a child.[49] This is directly relevant to the distinction we are making between ecocentric restoration and self-interested resource conservation. It is not that activities associated with resource conservation—gardening, for example, or examining soil profiles—cannot be undertaken as a focal practice; obviously they can. It is that *any* purpose beyond the act itself compromises its value *as* focal practice. Focus helps us reorient our lives through attention to what matters, Higgs notes. He points out that this is difficult "when we allow ourselves to be distracted by consumption,"[50] and he explores in some detail the prospects for the development of restoration as a focal practice.[51]

By taking restoration seriously as a context for the creation of meaning and value, restorationists would be returning to (in fact, restoring) an aspect of experience that has always played a crucial role in human life. Moreover, far from being of incidental importance, this is perhaps the most comprehensive and stringent context in which to consider the nature and import of *any* act—not just in terms of its product but in terms of the meanings that can be found in or coaxed from it. Meanings depend on distinctions and emerge from acts that self-consciously enact these distinctions, dramatizing them in psychologically effective ways.

Ecocentric restoration, for example, is a dramatic—that is, active—response to the experience as an "other" of nature and, linked to that, its intrinsic value. Indeed, it is in just this "fourth dimension" of value we call ritual or performance that the distinction between holistic and meliorative, self-interested and other-regarding land management is felt most strongly and is most clearly and powerfully experienced and expressed. And although some may object that this is impractical or merely fanciful, the record of place-based peoples who have inhabited ecosystems for millennia, depending on make-believe—that is, ritual—as a basis for land management forces us to admit that this "soft" or "subjective" approach to land management actually works better than purely technical or literal practice.[52] The reason for this is obvious: The "soft" path defined by the technologies of myth, ritual, art, and religion deals with the subjective dimension—the values, beliefs, ideas, anxieties, and affections that define and enact the relationship between humans and their environment—in a way neither technique nor ideas alone can.

Impractical as it is, ecocentric restoration will never be the land management protocol of choice for most of the planet. Rather, we might expect it to be practiced on a modest scale in most areas in order to create and maintain old associations as emblems of the past—the "vignettes" of the Leopold Committee—and to provide habitat for those human-intolerant species Mike Rosenzweig calls "kulturmeidern."[53] But the restoration going on in these small areas may also serve, like theatrical, athletic, or religious performances, as occasions for confronting and coming to terms with our experience of the world and each other, including other species, at the deepest levels. In this way they may both express and reinforce the valuing of nature for its own sake that thinkers have urged for generations.

This "use" of restoration remains to be explored and tested systematically, however. In the meantime, restorationists go on doing what human beings have always done, inventing ways to enhance the performative aspects of their work, at times moving it up the scale of performative intensity to the level of ritual. Lake Forest, a suburb north of Chicago, for example, celebrates an annual festival held around the time of the autumn equinox and centered around the burning of the large brush piles created by restorationists clearing buckthorn and other exotics from oak openings.[54]

In a similar move from the purely practical into the domain of the expressive and fantastic, the crew at Glacial Park, a 1,280-hectare "restoration park" 80 kilometers west of Chicago, has taken to scheduling night burns on its prairie and savanna restoration sites. Bill Jordan asked crew boss Brad Woodson why they did this, with the inconvenience of working nights, the hassle of planning a burn for the benefit of an audience, and the increased likelihood of losing tools or a crew member twisting an ankle in the dark. "Because it's spectacular," he said, "and people love it."[55]

Mark Stemen, who teaches a hands-on course on restoration at California State University–Chico, has integrated low-key ritual into the program, in effect sneaking up on the puritans in class. More formally, the community that has been restoring salmon habitat in the watershed of the Mattole River in Northern California for more than three decades has turned its work into the subject of a musical comedy in which, a reviewer wrote, "loggers and hippies, backpackers and ranchers, biologists and businessmen—and even the notorious spotted owl—represent the trials and tribulations of a community in the midst of a resource war."[56]

And at the other end of the seriousness spectrum, perhaps the closest approach to the high-end aims of ritual in providing a context for confronting the existential challenges of birth, killing, death, and mortality we are aware of is the salmon reproduction ceremony celebrated by the Mattole River restorationists. Explicitly grounded in respect for the practices of pre-Columbian inhabitants of the region and eloquently described by Freeman House,[57] it begins with the killing of the mother fish:

> Each of us has performed this rite a number of times before, but it never ceases to be weighted with nearly intolerable significance, the irreducible requirement to do it right.
>
> I have handed the ironwood club to Stevie. . . . He is coiled tightly like a baseball batter at plate; he squints at the fish with his one good eye. It will be a mutual embarrassment if it takes him more than one blow to kill her. The club comes around . . . and connects solidly at the back of her head, just behind her eyes. She shudders for a moment and is still. Stevie drops the club in the grass.
>
> "Good," Gary murmurs, "good."

Then the fertilization:

> I lower the fingers of one hand into the heart of creation and stir it once, twice. For a moment my mind is completely still. Am I holding my breath? I am held in the thrall of a larger sensuality that extends beyond the flesh.
>
> By the time I have lifted my fingers out of the bucket, fertilization will have taken place or not.

Chapter 11

Current Thinking

As its history makes clear, ecocentric restoration is an elusive idea that entails troubling contradictions and ambiguities, challenging not only the land manager but also the environmental philosopher. The result, as restoration has gained importance both as a conservation strategy and as a buzzword, has been a fascinating discussion, at times rising to the temperature of a debate, over the ends and means of restoration. Here we conclude our story of the development of this idea with an overview of this discussion during the past decade or so.

First, as far as debate about the importance of ecocentric restoration in preserving natural ecosystems is concerned, the argument is pretty much over. Although a few still harbor misgivings, most land managers and environmentalists accept the idea that this version of restoration is not just an emergency measure or a second-rate alternative to preservation but the necessary (if imperfect) means for achieving it. Similarly, although ideas about what restoration is vary widely, the idea—or ideal—of ecocentric restoration has become an important source of energy driving environmental thinking and practice. Projects defined or inspired by this idea have proliferated in almost every part of the world, providing, among other benefits, occasions for discussion and debate about the questions they raise. For example, when Pauline Drobney says of the restoration effort at Neal Smith National Wildlife Refuge in Iowa that "our aim is to get back as many pieces as we can," including the earthworms,[1] we recognize a commitment that is evident in projects at countless other sites around the world.

At the same time, there is lively debate about the idea of the historical model that defines ecocentric restoration and distinguishes it from other forms of restorative land management. *Restoration* obviously means bringing something back into a previous condition; the question is, What condition? And why? The early projects we have identified as startup examples of ecocentric restoration may have been motivated partly by bioaltruism but were also justified by the notion that the ecosystems they aimed to replicate were models of land health and so are exemplars in some sense of an ideal habitat for our own species. However, this notion was based in part on the idea that certain ecological communities are ecologically privileged, climax communities with special qualities of integrity and stability, an idea that no longer carries much weight among ecologists. For some restoration skeptics this has meant that the choice of a model for a restoration effort is arbitrary or merely a matter of self-interest. These critics insist that ecosystems that happen to have existed at some time in the past, even for long periods, may provide *examples* of associations that persisted under a particular set of circumstances, but they have little value as models of land health, especially under altered conditions. So writer Keith Kloor asks "whether restoring a forest to its presettlement condition is even a legitimate goal, considering that Native Americans were shaping the land long before European settlers arrived."[2] This is interesting as a reflection of the Clementsian or edenic idea that restoration properly means a return to an "original" or "natural" condition, with *natural* being understood as "without human influence," so that an attempt to return an ecosystem to a merely *historic* condition, perhaps reflecting human influence, somehow delegitimizes it.

History, of course, is messy in ways that myth and idealized ecosystems are not, and it offers a less enchanting model for restoration efforts, especially when it turns out that the old ecosystems are in many ways ecologically and economically obsolete, so that attempts to restore them fall somewhere between impractical and quixotic. This raises an important question: Why try to return an ecosystem—or for that matter anything—to some previous condition unless that condition is in some way better than its present condition?

Restorationists have given this matter a good deal of thought in recent years and have articulated ideas ranging from skepticism about the value of the historic model to serious commitment to it. Speaking at the Society for Ecological Restoration (SER) conference in Perth in 2009, Don Falk, a former director of the society who is now with the Laboratory of Tree Ring Research at the University of Arizona, provided a good example of this thoughtful head-scratching. Falk acknowledged that adoption of a historic "reference system" is what distinguishes restoration from other forms of land management and suggested that this has special value as a governor on human imagination and pride, but he characterized a too-literal adoption of the historic model as a "cartoon" version of restoration that ignores the realities of evolution and ecological change. Others take a less ambiguous view, regarding the historic system as a point of departure rather than an actual model. Mark Davis and Lawrence Slobodkin have suggested that "restoration ecology should look primarily to the future when defining its goals"[3] (raising the question of why it would be called "restoration"), and they propose a definition of *restoration* that abandons the historic model entirely, making it "the process of restoring one or more valued processes or attributes of a landscape."[4] (This, of course, is what used to be called conservation and leaves no room at all for ecocentric restoration.) Wetland restorationist Ed Garbisch rejected the idea of dedication to a historic model long ago, as we have noted. Steve Packard says that he regards the historic ecosystem not as a literal model for the restorationist but as a "metaphor" for a healthy, biotically rich, perhaps "natural" ecosystem. And Ed Collins, who conceived the idea of turning Glacial Park, west of Chicago, into a "restoration park" and spearheaded the re-meandering of a 2.5-kilometer stretch of river in the park, including reconstruction of two large glacial kames,[5] writes,

> My opinion is that restoration is not about historic landscapes at all. I think that's a lost cause. I have always felt it is about a new landscape for a new type of society. I guess that's why I never get too riled up over the genotype and pre-settlement debate. It's about the next phase of society, not one that is long gone.[6]

This emphatic rejection of historic ecosystems as models for restoration is not universal among those involved in restoration efforts, however. Anthony Bradshaw, a British ecologist and early leader in the field of restoration, noted in a 1995 article that the word *restoration* denotes a return to a previous condition regarded as "perfect" (that is, complete) and that this distinguishes it from other forms of land management.[7] While acknowledging the "troublesome perfectionist implications" of the word, he drew approvingly on a definition published by a panel chaired by John Cairns Jr. to argue that even though the practitioner may focus on selected features of an ecosystem for one reason or another, "our ultimate aim should be restoration of the whole ecosystem."[8] In a similar vein, James Aronson and several of his colleagues, while acknowledging that restoration goals should accommodate a range of developmental trajectories rather than just one, have stressed the importance of reference systems, noting that overemphasis on the dynamic character of ecosystems is likely to weaken arguments for conservation of historic systems.[9] Environmental philosopher Eric Higgs argues that restoration will have "lost its way" if restorationists focus on "services" at the expense of historical continuity, sense of place, and engagement with nature. In his 2003 book *Nature by Design*, Higgs explores the tension between respect for history and the challenge of responding to altered ecological and social conditions. Itemizing the values people find in historic systems as nostalgia, story, and sense of deep time, all of which underlie sense of place, he makes a case for what he calls "fidelity" to historic models, a term that he prefers because it implies a more dynamic conception of the model system than terms such as *baseline* or *benchmark*, which he thinks convey a static, "snapshot" idea of the model system.[10] And providing a technical foundation for fidelity of this kind, Dave Egan and Evelyn Howell have put together what amounts to a catalogue of ways to recover historical information as a first step in the restoration of historic ecosystems.[11]

Debate over ways of construing the idea of restoration was evident at a symposium on the role of the historic ecosystem in restoration, held in Zurich in 2006. As participant T. C. Smout pointed out, the discussion included a number of protocols under labels such as *rewilding, regardening, conservation gardening*, and *landscape-scale change*, each of which has its own value. This is exactly what we have in mind when we insist on the distinction between ecocentric restoration and meliorative land management. What is intriguing about this conversation from our perspective is not so much the dismissal of the "old" idea of restoration of whole ecosystems but rather how that idea persists, serving as a kind of promise—or challenge or threat—around which the conversation is organized.

Why?

Turning to the question of motives—why do people set out to restore an ecosystem in the first place?—we find a similar ambivalence and variety of ideas. This conversation sounds to us like an ongoing attempt to justify ecocentric restoration in terms of self-interest. This may be effective in a political or even ecological sense, but it deprives those involved, including the audience of observers and taxpayers, of the experience of ecocentric restoration and the distinctive values that may emerge from it. In any event, it is probably fair to say that concern for the long-term conservation of species and ecosystems is a serious concern of most of those involved in the restoration culture that has emerged from the history we are writing

Extreme Restoration

Although some have questioned the classic notion of restoration as an attempt to return to an idealized past that overlooks both the dynamic nature of ecological systems and the influence of indigenous peoples on their environment, restorationists have been busy breaking out of the old box of 1491 (or its regional equivalent) while exploring the variations that can be played on the theme of ecocentric restoration.

Since 1995, for example, managers in New Zealand have been working on restoring a 225-hectare site on the outskirts of Wellington to the condition it was in at the time of first human contact, some eight centuries ago. The project, at the Karori Sanctuary, has involved construction of an 8.6-kilometer predator-proof fence, extirpation of thirteen species of mammals inside the exclosure, and reintroduction of a number of native mammals as part of a five-hundred-year plan to restore lowland forest and wetlands thought to be native—really native—to the site.[a] Obviously wildly impractical, the project exemplifies the value of restoring an ecosystem to the condition it was in when we found it—*we* in this case meaning not just Europeans but human beings.

In the meantime, a number of ecologists have proposed versions of "Pleistocene rewilding," proposing the use of large mammals still existing in Africa and Asia as surrogates for the mammoths, mastodons, camels, and other megafauna that disappeared in most parts of the world at the end of the Pleistocene, some thirteen thousand years ago. Since the 1980s, ecologist Frans Vera has been supervising reintroduction of deer, wild cattle, and horses at Oostvaardersplassen, a 6,000-hectare reserve outside Amsterdam, in part to study the role of large animals and humans in community dynamics.[b]

Around the same time, Sergey Zimov and his colleagues, working in the Russian republic of Yakutia, launched Pleistocene Park, aiming to reintroduce surviving species of large mammals as part of an attempt to restore the taiga and tundra of the region, determine the role animals played in maintaining those ecosystems, and test the idea that hunting by humans played a role in the extinction of megafauna at the end of the Pleistocene.[c] Ecologist Michael Soulé proposed a similar project for North America in a speech to the Society for Conservation Biology in 1990.[d] And in 2005 Cornell University ecologist Josh Donlan and his associates proposed the imaginative strategy of using closely related surviving species as proxies for extinct large vertebrates in some ecosystems, in part to provide refugia for megafauna threatened in their native habitats in Africa and Asia.[e]

Extreme Restoration
Continued

These visions of a future deeply informed by a long-lost past and dreamed up by ecologists may seem like science fiction along the lines of *Jurassic Park*, but they intersect with practical and economic considerations in the proposal by Deborah and Frank Popper for creation of a "Buffalo Commons," entailing the return of bison to the American West on a subcontinental scale.[f]

At the same time, pursuing the "logic of restoration" in other directions, restorationists have launched projects that exemplify the notion of ecocentric restoration even as they defy the notion of restoration as the attempt to re-create "natural" conditions, emulate an ideal ecosystem, or improve human habitat. These include attempts to restore economically obsolete agricultural landscapes in various parts of the world,[g] restoration of vegetation at battlefields and other historic sites, and even a project to restore the ecology of a Nazi concentration camp by removing the ruderal vegetation that had developed there during the decades since the end of World War II.[h]

---

a. www.sanctuary.org.nz/Site/Conservation_and_Research/Restoration /The_fence.aspx. For a discussion of ideas about restoration of prehuman ecosystems in New Zealand developed by paleontologist Matt McGlone, see Lesley Head, *Cultural Landscapes and Environmental Change* (London: Arnold, 2000), 103–5.

b. Andrew Curry, "Where the Wild Things Are," *Discover* (March 2010): 58–65.

c. Sergey A. Zimov, "Pleistocene Park: Return of the Mammoth's Ecosystem," *Science* 308 (May 2005); www.sciencemag.org/cgi/content/full/308/5723 /796.

d. Michael A. Soulé, "The Onslaught of Alien Species, and Other Challenges in the Coming Decades," *Conservation Biology* 4 (September 1990): 235; Paul S. Martin and David A. Burney, "Bring Back the Elephants," *Whole Earth* (Spring 2000), http://www.wholeearth.com/issue/2100/article/16/bring.back.the .elephants#content (accessed March 12, 2011).

e. Josh Donlan et al., "Re-wilding North America," *Nature* 436, no. 18 (August 2005): 913.

f. Anne Matthews, *Where the Buffalo Roam: The Storm over the Revolutionary Plan to Restore America's Great Plains* (New York: Grove, 1992).

g. See Peter A. Bowler, "In Defense of Disturbed Land," *Ecological Restoration* 10, no. 2 (1992): 144–49; Ikuyo Okada, "Restoration and Management of Coppices in Japan," *Ecological Restoration* 17, no. 1–2 (1999): 31–38.

h. William R. Jordan III, "Appellplatz," *Ecological Restoration* 15, no. 2 (1997): 115.

about here. At the same time, as the experience of advocates for the environment makes quite clear, appeals to bio-altruism and the intrinsic value of small fish, obscure flowers, and even panda bears go only so far. Recognizing this, advocates for restoration projects often emphasize self-interest, much as Aldo Leopold did in making the case for the restoration project at the UW–Madison Arboretum three-quarters of a century ago. Because the interests of people and old ecosystems often broadly overlap, especially when human interests are considered broadly to include the aesthetic, psychological, moral, and spiritual dimensions of value, this two-pronged argument does not necessarily compromise the aims of ecocentric restoration. At the same time, because they do not *necessarily* coincide, merely conflating them, downplaying the distinction between self-interest and the well-being of the ecosystem, or dismissing it as meaningless,[12] ultimately endangers the elements of "old" nature that do not happen to fall within the circle of what a particular culture happens to need, want, or value.

Some, concerned about this, have advanced both philosophical and ecological arguments against the foregrounding of self-interest at the expense of commitment to the ecosystem itself. Conservationists such as Reed Noss and Michael Scott argue outright for large-scale restoration on behalf of what amounts to the "old" aim of preservation.[13] And Michael J. Stevenson at the School of Forestry and Environmental Studies at Yale University, while acknowledging that "an economic approach" may be "politically expedient" in making the case for restoration, finds the argument for making economic self-interest a "primary rationale for restoration" "dangerous" and argues that "to wholeheartedly embrace such an approach to the detriment of 'humanistic, psychological, or biological reasons' for restoration is a moral failure that threatens to tear at the fabric of restoration ecology." He points to the Endangered Species Act as evidence that "modern society has reached a point where it has the potential to transcend the limits of economic approaches to conservation."[14]

This is not an isolated position. Historian David Lowenthal notes that when some three hundred river restorationists and stakeholders in thirty-six countries were asked in a 2003–04 survey what they meant when they used the term *restoration*, four out of five chose the strict definition proposed by John Cairns Jr. in 1991—"complete structural and functional return to a predisturbance state"—even while acknowledging that they rarely achieved this goal.[15]

Others, however, recognizing the importance of appealing to a wide constituency and unwilling to rely on appeals to bio-altruism to make their case, emphasize the value restoration may have as an investment in natural capital in the form of ecosystems that provide goods such as food, fuel and fiber, and services such as flood control and sequestration of carbon dioxide. Under this rubric, as described by James Aronson and several of his colleagues, restoration is an investment in nature motivated by self-interest, broadly understood not merely in economic terms but taking into account "all aspects of human well-being" (5).[16]

Of course, this raises questions about the fate of the old ecosystems, keeping in mind that humans have been tinkering with and "improving" the ecosystems they inhabit for as long as we have any record and that a culture may not perceive a particular old ecosystem as entirely consistent with its interests, however broadly conceived. Addressing this question, Aronson and his colleagues acknowledge that an emphasis on human interests may obscure the idea that ecosystems and species have intrinsic value and "are worth restoring and preserving 'for their own sake,' regardless of their economic (or other) value to humans." However, they argue that it is necessary to emphasize self-interest in order to "mainstream ecological restoration into the economy," and they suggest that, in any case, ecosystems modeled on historic systems, to the extent that they are "self-supporting," "will adapt to climate change and evolve as well as or better than 'designer' ecosystems" (7).

In a similar vein, attempting to close the gap between bio-altruism and self-interest, some restorationists have used the argument that an ecosystem needs "all its parts" to function properly. Thus Andy Clewell argues that although there are obviously many legitimate motives for restoration efforts, the "pragmatic" one of restoring natural capital must be "the primary rationale for restoration in most regions of the world."[17] Having asserted this, however, Clewell goes on to argue that the emphasis on human interests actually entails a commitment to the restoration of whole systems, and he cites Aldo Leopold's argument that unless all the species are included, "the restored ecosystem may not regain its former structure and may not function as well as it once did."[18]

The idea, related as it is to the idea that nature constitutes a Great Chain of Being with an integrity conferred on it by the Creator, that there is a close relationship between the species composition and the dynamics of an ecological system is intuitively appealing, and advocates for the conservation of species and biodiversity have long used it to make a utilitarian case for the conservation of economically "useless" species. The problem with this argument is that, handy as it may be, ecologists no longer believe that there is a straightforward relationship between the species composition or biodiversity of an ecosystem and its stability or productivity. Mark Schwartz, an ecologist at UC–Davis, notes that although recent research in this area does indicate a positive relationship between diversity and various measures of ecosystem function, the relationship is complicated. Some species play much more important roles than others, and in systems in which the relationship between species diversity and function has been studied in detail, the curve levels off after relatively few species. This is ongoing research, but Schwartz says that the results at this point suggest that there is a high degree of species redundancy, as far as ecosystem function is concerned, and that in many cases most of the functions of an ecosystem are up and running with as few as 10 percent of the species that normally inhabit it.[19]

So there is a range of views with respect to motives and to how best to make the case for restoration. What is striking about the resulting conversation, however, is the "pragmatists'" attempt to justify concern for "all the parts" without appealing to their intrinsic value. These attempts to smuggle ecocentric restoration onto the agenda without calling it that, and without reference to the problems of otherness, conflicts of interest between us and other species, natural disharmony, the philosophical enigma of altruism, and the question of self-sacrifice, indicate clearly that those involved really do value the old ecosystems "for their own sake" but are aware that this argument carries little weight in a society that tends to regard self-interest as the mainspring of human behavior. Perhaps what we are seeing here is an international version of historian David Nirenberg's comment that "after decades of triumphant liberal capitalism, we lack—at least in the United States—a political or philosophical discourse of sufficient resonance to temper the claims of self-interest."[20]

It is interesting to consider in this connection the contrast between what we might call preservation-oriented and restoration-oriented arguments on behalf of conservation. Although appeals to the intrinsic value of species and ecosystems were a mainstay of the rhetoric of the preservation-oriented environmentalism of the 1960s and 1970s, those making the case for restoration seem less willing to advance this argument and more inclined to emphasize appeals to self-interest. Perhaps this is because restoration is a more demanding test of allegiance to the notion of intrinsic value. It is one thing to argue that something—snail darters, rattlesnakes, whales, rainforests—should be left alone, another to insist that resources be invested to restore them.

Climate Change

Because the resulting "no-analogue" conditions could invalidate the whole idea of returning an ecosystem to a previous condition, the prospect of global climate change falls like a downed tree across the conversation about restoration.[21] Ecological restoration has always been about compensating for novel influences on an old ecosystem — by, for example, returning fire to a grassland or forest that has been deprived of fire. But changes in climate great enough to disrupt the historic relationship between climate and edaphic and hydrological conditions on the ground would render the old ecosystems irredeemably obsolete and preclude their restoration, at least outside climate-controlled conservatories. It would also invalidate any claim that might be made for their value as models for the conservation of natural resources, except in a highly abstract way.[22]

Despite — or because of — this, restorationists have taken a lively interest in the challenge posed by climate change, and many are coping with questions about how to design "natural" ecosystems to anticipate changes in climate or even in some cases how to respond to changes that are already taking place.

At the UW–Madison Arboretum, for example, scientists are experimenting with various forms of adaptive restoration in order to identify cause-and-effect relationships and to inform future restoration and management efforts.[23] In California, Nathaniel Seavy and his associates rely on their experience restoring ecosystems to advance several strategies to accommodate climate change and build resilience into riparian ecosystems. These include modifying horticultural practices to anticipate future uncertainty by planting species that are associated with a wide range of moisture conditions; they also include policies that allow land owners to restore habitat for endangered species free of legal responsibilities for impacts during restoration and for maintaining such habitats indefinitely.[24] In the Pacific Northwest, Constance Millar and her colleagues argue that historic forests will not provide suitable models for restoration and management in an altered climate and suggest a "portfolio of approaches" to increase ecosystem resilience in response to climate changes.[25] They propose a number of adaptation and mitigation strategies based on the premise that no single solution will fit all future challenges and suggest that effective responses will include assisting species migration, creating "porous" landscapes, and allowing for genetic diversity in planting mixes.[26] In a similar vein, Peter Dunwiddie and his colleagues in Washington State discuss their strategies within three general categories — component redundancy, functional redundancy, and increased connectivity — while acknowledging that none of these strategies are failsafe or risk free.[27] And De-Paul University ecologist Liam Heneghan notes that there is much to learn about "the limits of compatibility" between ecosystems and that it may be possible to "reverse the flow" of influence so that restored ecosystems influence their surroundings as well as the other way around.[28]

Yet another step away from the classic idea of restoration as the restoration of "all the parts" is an ongoing, lively discussion of the idea of creating frankly novel ecosystems in response to or in anticipation of major changes in environmental conditions.[29] This strategy may fall short of restoration of all the parts, but it does not preclude attempts at ecocentric restoration — that is, restoration centered on the ecosystem, even when this entails compromising historical accuracy. That is simply what the restorationist who is committed to restoration of the whole ecosystem has to do in order to turn it edgewise, as it were, to get it through the knothole of drastic, effectively irreversible changes in conditions. This is not giving up on the old ecosystem. Rather, it is like jamming a piano through a door to rescue it from a burning building, something you do even if it might mean knocking off a leg or two and getting the instrument out of tune.

Conscience Change

All these are strategies for *responding* to human-driven climate change, however. Others have been exploring the role restoration might play in forestalling or at least minimizing it. So far, however, this work has been mostly technical in nature, involving mainly restoration — or invention — of ecosystems that absorb and sequester more carbon than other ecosystems do.[30] Much less attention has been paid to its value as a context for negotiating the relationship between people and their environment. Many practitioners appreciate the value of restoration as a way of "reinhabiting" natural landscapes, and some have even exploited this value through the development of restoration as a performing art, as we noted in chapter 10. Progress in this crucial "fourth dimension" of value has been limited, however. This is hardly surprising in a culture that has relegated responsibility for ecological restoration to scientists and has little facility for deploying the technologies of value creation, meaning making, and conscience formation in the pubic arena.

Illinois restorationist Tom Simpson argues that what we are seeing here is the result of the "capture" of restoration by ecology and its subdisciplines, a development that has limited realization of its value in the dimensions of experience and value creation. Although other disciplines — history, philosophy, sociology, and a few others — have made contributions, their role has been unmistakably second-fiddle, and the skepticism about the efficacy of performance that characterizes the culture of science ensures that an entire dimension of experience is hardly repre-sented at all.[31] Performance, which has played key roles in human rela-tionships for millennia, has played, at best, a submarginal role in the de-velopment of restoration and realization of its values.

Seen in this limited, one-dimensional way, restoration may seem a small thing indeed and therefore easily dismissed. For example, sociologist Jan Dizard writes that the problem with restoration is "as simple as it is painful: each project asks us to return to an idealized past that cannot be recaptured on a scale that meaningfully addresses our very real environ-mental woes."[32] Characterizing the aims of restoration as a quixotic at-tempt to return to an "idealized"—that is, imaginary—past is a common device for dismissing it in order to replace it with something else. What Dizard fails to take into account is the value a restoration effort may have as an experience and an expressive, value-creating or conscience-forming act. Parts of things—symbols, gestures, relics—are part of the grammar of these dimensions of human life. And it is arguably by way of just such small, symbolic projects, which reduce effective work to an expressive act, that people negotiate the inner transformation of mind and spirit on which the fate of the world ultimately depends.

In any case, to the extent that modern societies can find effective ways to exploit these ancient technologies of imagination and conscience, this lends weight to Tom Simpson's suggestion that the best way not only to re-spond to the prospect of climate change but to forestall it "is to restore eco-systems now."[33]

Part and Apart

As the notion of ecocentric restoration—or perhaps more accurately the word *restoration*—has spread beyond its cradle of origin in North America and Australia to parts of the world lacking the "discovery" experience pe-culiar to these parts of the world, there has been a tendency to construe it in a broad sense to include a wide range of land management protocols that a few decades ago would have been called conservation. This broad-ening of the idea is often defended on the grounds that restoration of old (and, as we have noted, often economically as well as ecologically obso-lete) ecosystems is not only irrelevant in Old World or developing world settings but is a luxury only a rich society can afford.[34]

This contention brings into the open a question that has been implicit in the restoration effort ever since the first tinkerers decided to try to re-store whole ecosystems rather than selected features of ecosystems in

The Critics

As restoration, broadly conceived, has gained stature as an item in the conservation repertory, it naturally has attracted the attention of critics who find fault with it in various ways. In 1982, even before he was aware that people were attempting to restore "natural" ecosystems, Robert Elliot, an Australian environmental philosopher, proposed it as a kind of thought experiment, which he then attacked in an article provocatively titled "Faking Nature." In this article Elliot critiqued what he called the "restoration thesis—the idea that a restored thing has the same value as the "original,"[a] an idea he did not, it should be noted, attribute to any actual restorationist or school of restorationists. Environmental philosopher Eric Katz has developed this critique further, objecting to what he sees as the human-centered emphasis he finds in even the most ostensibly ecocentric projects, exemplified by the restoration efforts under way in Chicago's Forest Preserves.[b]

In a 1994 article that aroused some consternation in restoration circles, Michael Pollan, who is widely known for his writing on food and agriculture, published an essay questioning the exclusion of nonnative plants from gardens and landscapes, citing work by landscape historian Joaquim Woelschke-Bulmahn, who argued that the idealization of native plants by the Nazis was the horticultural version of their campaign against non-Aryans.[c]

More recently, sociologist Jan Dizard has argued that the decline of the idea of the stable climax ecosystem deprives the old ecosystems of any special claims as models, especially considering that ideas about what is desirable vary widely.[d] This, of course, overlooks the possibility that there may be some value in restoring and maintaining *undesirable* ecosystems.

And a group at the University of Maryland recently published a list of a half-dozen ideas they characterized as "myths" that they regard as prevalent among restorationists, which, they argue, can lead to "conflict and disappointing results, if not recognized as myths." Central to these is what they call "the myth of the carbon copy"—that is, the idea that ecosystems *can* be restored, which they suggest reflects a general failure to recognize the variability of ecosystems and hence the uncertainties associated with attempts to restore them. Consistent with the move toward more abstractly defined objectives that, as we have seen, has been an element in thinking about ecocentric restoration from the beginning, they argue for more emphasis on resilience and adaptive management.[e]

The Critics
Continued

a. Robert Elliot, "Faking Nature," *Inquiry* 25 (1982): 81–93; also Robert Elliot, *Faking Nature: The Ethics of Environmental Restoration* (London: Routledge, 1997).

b. Versions of "The Big Lie," Katz's early paper on this topic, appeared in *Research in Philosophy and Technology* 12 (1992); in *Restoration and Management Notes* 9, no. 2 (1991): 90–96; and in Katz's collection of essays *Nature as Subject: Human Obligation and Natural Community* (Lanham, MD: Rowan & Littlefield, 1997), 93–107. For a discussion of these ideas, see Eric Higgs, *Nature by Design: People, Natural Process and Ecological Restoration* (Cambridge, MA: MIT Press, 2003), 218–21.

c. Joaquim Woelschke-Bulmahn, "Some Notes on the Mania for Native Plants in Germany," *Landscape Journal* 1, no. 2 (1992): 116–26; "The 'Wild Garden' and the 'Nature Garden': Aspects of the Garden Ideology of William Robinson and Willy Lange," *Journal of Garden History* 12, no. 3 (1992): 183–206; Michael Pollan, "Against Nativism," *New York Times* Magazine, May 15, 1994. Bill Jordan responds to this article in "The Nazi Connection," *Restoration & Management Notes* 12, no. 2: 113. Pollan explores the idea of restoration sympathetically in his *Second Nature: A Gardener's Education* (New York: Atlantic Monthly Press, 1991), see especially chapter 10.

d. Jan Dizard, "Uneasy Relationship between Ecology, History, and Restoration," in Marcus Hall, ed., *Restoration and History: The Search for a Useable Environmental Past* (New York: Routledge, 2010), 154–63.

e. R. H. Hilderbrand, A. C. Watts, and A. M. Randle, "The Myths of Restoration Ecology," *Ecology and Society* 10, no. 1 (2005), www.ecologyandsociety.org/vol10/iss1/art19/.

which they happen to have a special interest. Given that restoration of ecosystems often serves human ends, what happens on the occasions when it does not—when it turns out that altruism is not in our self-interest?

This is not a question peculiar to the business of restoration. It arises from every kind of relationship: Am I being nice to this guy because I think he might lend me his car or offer me a job someday—or what? However, it is thrown into relief by the mere *act* of attempting to restore an ecosystem with all its parts, and the claim—or pretense—of altruism that might be read into such an act.

This may be the most important thing about ecocentric restoration: It raises this basic question about ourselves and our motives in a way that, say, growing cabbages does not. Wrestling with this question does not lead to definitive answers, but it does do us good. This is evident in the widely cited definition of *ecological restoration* promulgated by the SER in its *Primer on Ecological Restoration*: "Ecological restoration is the process of assisting the recovery of an ecosystem that has been degraded, damaged or destroyed."[35] Hammered out over a period of more than a decade by a consortium of SER members representing a wide range of interests and points of view, this definition, together with the commentary that accompanies it, represents a systematic attempt to come to terms in a practical way with the ambiguities inherent in the notion of restoration, especially when it is applied to a dynamic entity such as an ecosystem. These include the question of how seriously—or literally—to take the idea of the historic ecosystem as a model, how to choose and define models, and what to make of the human role in the shaping of ecosystems such as the prairies of the American Midwest or, for that matter, the Amazonian forests, which used to be regarded as natural but now are understood to have been shaped in large part by human beings.

They also include the questions of motives, of self-interest and eco-altruism that underlie the distinction we have made between ecocentric restoration and self-interested forms of land management. Certainly, if restoration is understood, as characterized in the *Primer*, as an attempt "to return an ecosystem to its historic trajectory," this is an aim that will often run counter to the wishes and, we must suppose, the interests of those who inhabit it. Implicitly acknowledging this, the *Primer* hedges on a strict commitment to historical authenticity, allowing for compromises with respect to exotic species, for example, and for the inclusion of self-interested objectives such as the enhancement of natural capital and "aesthetic amenities," even when they are not integral to the nine attributes the *Primer* identifies as attributes of a restored ecosystem.

The result is a broad, inclusive—we might say lenient—idea of restoration, tailored to allow a wide range of land management practices to claim the rubric of "restoration," currently a bestseller in conservation circles. What it does not account for is the disparity that often exists between the interests of an old ecosystem such as a tallgrass prairie and those of a contemporary human community living in, say, a suburb of Milwaukee, or the costs and sacrifices the restoration and management of such an ecosystem might entail.

* * *

Does this matter? After all, humans are radically dependent on nature. The vision of an untroubled relationship with it is certainly an appealing one, and the emphasis on membership in the land community that has characterized the environmentalism of the past few generations is certainly an improvement on the notion that nature is nothing but a repository of resources. At the same time, human relations with the rest of nature, like the relationships between other species, have never been trouble-free. Tensions exist between nations and tribes, between individuals, between self and other, and between human cultures and the rest of nature—tensions that, unless we are creationists, we have to suppose are themselves natural.

Ecocentric restoration is important not because it is better than other forms of land management in an ecological or moral sense but because it complements self-interested forms of land management, providing, in a way they do not, a context for exploring and paying tribute to the rest of nature even when its interests do not coincide with our own.

In other words, it poses, at the level of the ecosystem, the question of our relationship with a nature with which we sometimes find ourselves at odds, that is perhaps at odds with itself, and that, as Emerson wrote, "leads us on and on, but arrives nowhere; keeps no faith with us."[36]

The question is—and always has been—what do we do, and how do we come to terms with a nature like that?

This may be the most important thing about ecocentric restoration. It delivers us again and again to this question.

Introduction

1. Stephen J. Pyne, *Vestal Fire: An Environmental History, Told through Fire, of Europe and Europe's Encounter with the World* (Seattle: University of Washington Press, 1997), 27ff.

2. Marcus Hall, *Earth Repair: A Transatlantic History of Environmental Restoration* (Charlottesville: University of Virginia Press, 2005). For Hall's discussion of the deep history of restoration, see pp. 217–23.

3. Adapted from William R. Jordan III, *The Sunflower Forest: Ecological Restoration and the New Communion with Nature* (Berkeley: University of California Press, 2003), 22.

4. Theodore M. Sperry, "Reflections on the U.W. Arboretum 45 Years Later." Uncatalogued typescript in the arboretum's archives, dated February 10, 1982.

5. The word *we* is a technical term here, referring to individuals, groups of people, whole nations, or even the human race itself, who have brought about the alterations in an ecosystem that the restorationist aims to reverse.

6. Marcus Hall, *Nature's Repair: A Transatlantic History of Environmental Restoration* (Charlottesville: University of Virginia Press, 2005), 10, 223.

7. Jordan, *Sunflower Forest*, 85–87.

8. See, for example, Elizabeth Willott, "Restoring Nature, without Mosquitoes?" *Restoration Ecology* 12, no. 2 (2004): 147–53.

9. Luke 10:29–37. Although we devised this formula ourselves, we were delighted to discover that it is consistent with what Michael Rosenzweig tells us is "a classic bit of rabbinic principle," for which he provided the following reference: Rabbi Yehuda said, The Rav said, "A person should busy himself with Torah and commandments even though not for their own sake because from 'not for their own sake' comes 'for their own sake.'" *Talmud Bavli, Tractate Horayot*, chapter 3, p. 10, column 2.

10. Jordan, *Sunflower Forest*, 20–21.

11. "Bruce Babbitt, Cornelius Amory Pugsley National Level Award, 2000," www.aapra.org/Pugsley/BabbittBruce.html (accessed February 14, 2011).

12. Edward O. Wilson, *The Diversity of Life* (Cambridge, MA: Belknap Press, 1992), 340.

13. Aldo Leopold, "The Land Ethic," in *A Sand County Almanac, with Essays on Conservation from Round River* (New York: Ballantine, 1966), 237–64. The phrase quoted is on p. 246.

14. L. E. Frelich et al., "Earthworm Invasion into Previously Earthworm-Free Temperate and Boreal Forests," *Biological Invasions* 8 (2006): 1235–45.

15. Roy A. Rappaport, *Ritual and Religion in the Making of Humanity* (Cambridge: Cambridge University Press, 1999), 86–97.

16. A. G. Tansley, "The Use and Abuse of Vegetational Terms and Concepts," *Ecology* 16 (1935): 284–307.

Chapter 1: Deep History

1. Curt Meine, "Conservation Movement, Historical," in *The Encyclopedia of Biodiversity*, ed. Simon Asher Levin (San Diego: Academic Press, 2001), 883–96. Also see P. S. Martin, "Pleistocene Ecology and Biogeography of North America," in *Zoogeography*, ed. C. L. Hubbs (Washington, DC: American Association for the Advancement of Science Publication 51, 1958), 375–420; P. S. Martin, "Prehistoric Overkill," in *Pleistocene Extinctions: The Search for a Cause*, ed. P. S. Martin and H. E. Wright Jr. (New Haven, CT: Yale University Press, 1967); and Paul Martin, "Prehistoric Overkill: The Global Model," in *Quaternary Extinctions: A Prehistoric Revolution*, ed. Paul S. Martin and Richard G. Klein (Tucson: University of Arizona Press, 1984), 354–403.

2. See Robert P. Edgerton, *Sick Societies: Challenging the Myth of Primitive Harmony* (New York: The Free Press, 1992).

3. Fikret Berkes, *Sacred Ecology*, 2d ed. (New York: Taylor & Francis, 2008). The material quoted is on p. 42 and is from V. H. Heywood, ed., *Global Biodiversity Assessment* (Cambridge: Cambridge University Press, 1995).

4. Ibid., 237. The reference to Eugene Hunn's work is on p. 53.

5. J. Donald Hughes, *Ecology in Ancient Civilizations* (Albuquerque: University of New Mexico Press, 1975), 50–51, 88–89.

6. M. Jha et al., "Status of Orans (Sacred Groves) in Peepasar and Khejarli Villages in Rajasthan" in *Conserving the Sacred for Biodiversity Management*, ed. P. S. Ramakrishnan, K. G. Saxena, and U. M. Chandrashekara (Enfield, NH: Scientific Publishers, 1998), 263–75. Narayan Desai, contact for the India chapter of the Society for Ecological Restoration, has pointed out that because Indian people value sacred groves for religious reasons, they are generally indifferent to arguments for their conservation based purely on an economic, which is to say distinctly modern, idea of value (personal communication, November 2005). In *Walden* Thoreau refers to the sacred groves of antiquity and the rituals associated with their management when arguing that the forests of New England have value beyond their value as resources, urging farmers who consider cutting a forest to recognize that it is "sacred to some god." *Walden and Other Writings of Henry David Thoreau* (New York: Modern Library, 1937), 225.

7. Hughes and Chandran, 69–86; G. Michaloud and S. Dury, in Ramakrishnan et al., *Conserving the Sacred*, 129ff.

8. Calvin Martin, ed., *The American Indian and the Problem of History* (Oxford: Oxford University Press, 1987).

9. Tim Ingold, *The Perception of the Environment: Essays on Livelihood, Dwelling and Skill* (London: Routledge, 2000), 58.

10. Raymond Williams, *Problems in Materialism and Culture* (London: Verso, 1980), 75. Quoted in Marcus Hall, *Earth Repair: A Transatlantic History of Environmental Restoration* (Charlottesville: University of Virginia Press, 2005), 23. Hall points

out that an emphasis on "man . . . [as] a free moral agent working independently of nature" is integral to the work of pioneer conservationist George Perkins Marsh and is reflected in Marsh's insistence on the role of humans as both disturbers and restorers of what he thought of as nature's harmonies.

11. In *The Reenchantment of the World* (Ithaca, NY: Cornell University Press, 1981), Berman suggests that what he calls "original participation," lacking awareness of the subject–object distinction, was the way people typically experienced the world before modernity and the rise of science, which he regards as a catastrophic loss of meaning. Carolyn Merchant offers a similar critique of science in *The Death of Nature: Women, Ecology and the Scientific Revolution* (New York: Harper & Row, 1980).

12. Reinhold Niebuhr, *The Nature and Destiny of Man* (New York: Charles Scribner's Sons, 1951), 14.

13. Mary Douglas, *Purity and Danger: An Analysis of the Concepts of Pollution and Taboo* (London: Routledge, 1966).

14. Signe Howell, "Nature in Culture or Culture in Nature? Chewong Ideas of 'Humans' and Other Species," in *Nature and Society: Anthropological Perspectives*, ed. Philippe Descola and Gísli Pálsson (London: Routledge, 1996), 127–44. The material quoted is on pp. 141–42.

Donald Brown includes "self distinguished from other," territoriality, nepotism, and ethnocentrism in a list of behavioral traits anthropologists have identified in every culture studied, in R. A. Wilson and F. C. Keil, eds., *MIT Encyclopedia of Cognitive Science* (Cambridge, MA: MIT Press, 1999). Brown's list is reproduced in Steven Pinker, *The Blank Slate: The Modern Denial of Human Nature* (New York: Viking, 2002), 435–39.

15. Roy F. Ellen, "The Cognitive Geometry of Nature: A Cognitive Approach," in *Nature and Society: Anthropological Perspectives*, ed. Philippe Descola and Gísli Pálsson (London: Routledge, 1996), 103–23.

16. Peter Singer, *The Expanding Circle: Ethics and Sociobiology* (New York: Farrar, Straus & Giroux, 1981), 111ff.

17. Howard L. Harrod, *Renewing the World: Plains Indian Religion and Morality* (Tucson: University of Arizona Press, 1987), 139.

18. William N. Fenton, *The Great Law and the Longhouse: A Political History of the Iroquois Confederacy* (Norman: University of Oklahoma Press, 1998), 259–60.

19. Singer, *The Expanding Circle*, 113ff.

20. Herbert N. Schneidau, *Sacred Discontent: The Bible and Western Tradition* (Baton Rouge: Louisiana State University Press, 1976), xi–xiii, 11ff.

21. According to Robert Nisbet, their commitment to reason led ancient Greek philosophers and historians to a similar perspective, and he finds an early expression of this in the eighth-century BCE farmer–poet Hesiod, in whose *Works and Days*, Nisbet writes, "For the first time an author addresses himself consciously and deliberately to the social and political problems of his day. He presents a vigorous arraignment of things as they are, and also a program of reform." Robert Nisbet, *A History of the Idea of Progress* (New York: Basic Books, 1980), 16.

22. Alexandra Kogl and David K. Moore, "Equality: Overview," in *New Dictionary of the History of Ideas*, vol. 2, ed. Maryanne Cline Horowitz (Detroit: Charles Scribner's Sons, 2005), 694–701; George L. Abernathy, *The Idea of Equality: An Anthology* (Richmond, VA: John Knox Press, 1959).

23. Lynn White Jr., "The Historical Roots of Our Ecologic Crisis," *Science* 155 (March 10, 1967): 1203–7.

24. Theodore Hiebert, *The Yahwist's Landscape* (Minneapolis: Fortress Press,

2008), 113–14. Examples of scholarship that has explored the Judeo-Christian tradition from an environmental perspective include Dieter T. Hessel and Rosemary Radford Ruether, eds., *Christianity and Ecology: Seeking the Well-Being of Earth and Humans* (Cambridge, MA: Harvard University Press, 2000); Sallie McFague, *A New Climate for Theology: God, the World, and Global Warming* (Minneapolis: Fortress Press, 2008); Paul Santmire, *The Travail of Nature: The Ambiguous Ecological Promise of Christian Theology* (Minneapolis: Fortress Press, 1985); Max Oelschlaeger, *Caring for Creation: An Ecumenical Approach to the Environmental Crisis* (New Haven, CT: Yale University Press, 1994); Susan Power Bratton, *Christianity, Wilderness and Wildlife: The Original Desert Solitaire* (Scranton, PA: University of Scranton Press, 1993).

25. Eric Katz, "Judaism and the Ecological Crisis," in *Nature as Subject: Human Obligation and Natural Community*, ed. Eric Katz (Lanham, MD: Rowman & Littlefield, 1997), 205–20.

26. Nelson Glueck, *Hesed in the Bible* (Jersey City, NJ: KTAV Publishing House, 1975). We are grateful to Prof. Michael Rosenzweig for alerting us to the idea of *hesed* and to this reference. For a discussion of the development of the related idea of hospitality, or rule of behavior toward strangers, in the ancient West, see Ladislaus J. Bolchazy, *Hospitality in Antiquity: Livy's Concept of Its Humanizing Force* (Chicago: Ares, 1977).

27. Jacques Derrida, "Hospitality," in *Acts of Religion* (New York: Routledge, 2002), 363. In *A History of the World in 10-1/2 Chapters* (London: Cape, 1989), 12ff, Julian Barnes riffs on this idea, noting that Noah is enjoined to bring seven pairs of the "clean" (i.e., edible) animals onto the ark but only one pair of the "unclean" animals (Genesis 7:1) and suggesting that the edible animals were overrepresented because they would serve as food for those on the ark, but the unclean—or useless—animals were also included in list of those to be saved. We are indebted to our colleague Liam Heneghan for drawing our attention to this idea and both treatments of it.

28. Norman Wirzba, *The Paradise of God: Renewing Religion in an Ecological Age* (Oxford: Oxford University Press, 2003), 34ff.

29. Yet another provocative partial precedent for the idea of restoration in the biblical tradition is the idea of the remnant, as of a people lost in some way, whose recovery or return is foreseen, a theme developed in many of the books of both the Old and the New Testaments. See entry on "Remnant" in Katherine Doob Sackenfeld, ed., *The New Interpretive Dictionary of the Bible* (Nashville, TN: Abingdon Press, 2009).

30. Arthur W. H. Adkins, "Ethics and the Breakdown of the Cosmogony in Ancient Greece," in *Cosmogony and Ethical Order: New Studies in Comparative Ethics*, ed. Robin W. Lovin and Frank E. Reynolds (Chicago: University of Chicago Press, 1985), 279–309.

31. Clarence Glacken, *Traces on the Rhodian Shore: Nature and Culture in Western Thought from Ancient Times to the End of the Eighteenth Century* (Berkeley: University of California Press, 1967), 23–24.

32. Michael Grant, *History of Rome* (London: Weidenfeld and Nicolson, 1978), 92ff. We gratefully acknowledge James Knight for drawing our attention to this development in Roman history and its relevance to the ideas we are exploring here.

33. David Rhoads, "Children of Abraham, Children of God: Metaphorical Kinship in Paul's Letter to the Galatians," *Currents in Theology and Mission* (August 2004).

34. This is not to discount the tension between Judeo-Christian and Pagan cultures, which have resulted in oppression and violence, often on a large scale. However, this does not negate the importance of the *principle* of the worthiness of the unfamiliar

other that took shape in the context of this tradition and is reflected in its tendency toward cross-cultural syncretism.

35. Madona R. Adams, "Augustine and Love of the Environment," in *Thinking about the Environment: Our Debt to the Classical and Medieval Past*," ed. Thomas M. Robinson and Laura Westra (Lanham, MD: Lexington, 2002), 77–78.

36. Glacken, *Traces on the Rhodian Shore*, 198. The passage from Augustine is from *The City of God*, trans. Marcus Dods et al. (New York: Random House, 1950), XI, 16.

Chapter 2: Run-Up

1. Donald Worster, *A Passion for Nature: The Life of John Muir* (Oxford: Oxford University Press, 2009), 289ff.

2. I. G. Simmons, *An Environmental History of Great Britain from 10,000 Years Ago to the Present* (Edinburgh: Edinburgh University Press, 2001), 118; Clarence J. Glacken, *Traces on the Rhodean Shore: Nature and Culture in Western Thought from Ancient Times to the End of the Eighteenth Century* (Berkeley: University of California Press, 1967), 173–75; Emile Mâle, *The Gothic Image: Religious Art in France of the Thirteenth Century* (New York: Harper and Brothers, 1958), 153. Philosopher John Passmore places this idea within the Western tradition of stewardship, calling for mastery of nature that perfects rather than destroys or enslaves. John Passmore, *Man's Responsibility for Nature: Ecological Problems and Western Traditions* (London: Duckworth, 1974), 33.

3. Quoted in Glacken, *Traces*, 298–99.

4. Lynn White Jr., "Cultural Climes and Technological Advance in the Middle Ages," *Viator* 2 (1971): 200; David F. Noble, *The Religion of Technology: The Divinity of Man and the Spirit of Invention* (New York: Alfred A. Knopf, 1997), 12–14, 15–17.

5. Lynn White, "Natural Science and Naturalistic Art in the Middle Ages," *American Historical Review* 52 (1947): 432–33; Roger D. Sorrell, *St. Francis of Assisi and Nature: Tradition and Innovation in Western Christian Attitudes toward the Environment* (Oxford: Oxford University Press, 1988).

6. "It was only because he possessed nothing," art historian Kenneth Clark writes, "that St. Francis could feel sincerely a brotherhood with all created things, not only living creatures, but brother fire and sister wind." Kenneth Clark, *Civilization: A Personal View* (New York: Harper & Row, 1969), 78.

7. Susan Power Bratton, *Christianity, Wilderness, and Wildlife: The Original Desert Solitaire* (Scranton, PA: University of Scranton Press, 1993), 218; Noble, *Religion of Technology*, 43–44.

8. Howard Mansfield, *The Same Ax Twice: Restoration and Renewal in a Throwaway Age* (Hanover, NH: University Press of New England, 2000), 245–47.

9. Samuel Y. Edgerton Jr., *The Heritage of Giotto's Geometry: Art and Science on the Eve of the Scientific Revolution* (Ithaca, NY: Cornell University Press, 1991), 40.

10. The classic statement of this critique is Lynn White's "The Historical Roots of Our Environmental Crisis," *Science* 155, no. 10 (March 1967): 1203–7. White's thesis has been both influential and controversial and in recent years has been superseded by research that has revealed more nature-friendly aspects of the Abrahamic religious tradition. Nevertheless, the idea that Western alienation from nature is a distinctive failing of this civilization has been a major theme of the environmental thinking of the

past half century, notable examples being the work of Carolyn Merchant, Morris Berman, Max Oelschlaeger, and Paul Shepard.

11. R. G. Collingwood, *The Idea of Nature* (London: Oxford University Press, 1945), 96.

12. Arthur Koestler, *The Sleepwalkers* (New York: Grossett & Dunlap, 1959), 322ff; I. Bernard Cohen, *Revolution in Science* (Cambridge, MA: Harvard University Press, 1985), 117–18.

13. Quoted from Paolo Rossi, *The Birth of Modern Science* (Oxford: Blackwell, 2001), 126–28.

14. William Eamon, *Science and the Secrets of Nature: Books of Secrets and Medieval and Early Modern Culture* (Princeton, NJ: Princeton University Press, 1994), 310–11.

15. Samuel Eliot Morrison, *Admiral of the Ocean Sea: A Life of Christopher Columbus* (Boston: Little, Brown, 1942), 556; Christopher Columbus, *Four Voyages to the New World*, trans. R. H. Major (Gloucester, MA: Corinth, 1978), 135.

16. John Prest, *The Garden of Eden: The Botanic Garden and the Re-Creation of Paradise* (New Haven, CT: Yale University Press, 1981), 6–9, 43–44.

17. Tim Ingold, *The Perception of the Environment: Essays on Livelihood, Dwelling and Skill* (London: Routledge, 2000), 57.

18. Arthur O. Lovejoy, *The Great Chain of Being: A Study of the History of an Idea* (Cambridge, MA: Harvard University Press, 1933). Keith Thomas, *Man and the Natural World: Changing Attitudes in England, 1500–1800* (New York: Oxford University Press, 1996), 278.

19. Mark V. Barrow Jr., *Nature's Ghosts: Confronting Extinction from the Age of Jefferson to the Age of Ecology* (Chicago: University of Chicago Press, 2009).

20. John Hansen Mitchell, *The Wildest Place on Earth: Italian Gardens and the Invention of Wilderness* (Washington, DC: Counterpoint, 2001), 26.

21. Ibid., 93–97.

22. Edward Hyams, *Capability Brown and Humphrey Repton* (New York: Scribner's Sons, 1971), 1; Passmore, *Man's Responsibility for Nature*, 36.

23. Glacken, *Traces*, 121.

24. A. T. Grove and Oliver Rackham, *The Nature of Mediterranean Europe: An Ecological History* (New Haven, CT: Yale University Press, 2001), 8–9; see also J. R. McNeill, *The Mountains of the Mediterranean World* (Cambridge: Cambridge University Press, 1992). Ruined landscape theory is not without its critics, among them Grove and Rackham, who note that writers have naively attributed too much degradation to the theory and overlooked alternative explanations.

25. Richard Grove, "The Island and the History of Environmentalism: The Case of St. Vincent," in *Nature and Society in Historical Context*, ed. Mikuláš Teich, Roy Porter, and Bo Gustafsson (Cambridge: Cambridge University Press, 1997), 149–50. Grove also notes interventionist controls on Formosa and in New India.

26. Quoted in *Images or Shadows of Divine Things by Jonathan Edwards*, ed. Perry Miller (New Haven, CT: Yale University Press, 1948), 135–36; J. Baird Callicott and Michael P. Nelson, eds. *The Great New Wilderness Debate* (Athens: University of Georgia Press, 1998), 5.

27. William Bartram, *Travels through North and South Carolina, Georgia, East and West Florida, the Cherokee Country, the Extensive Territories of the Muscogulges, or Creek Confederacy, and the Country of the Chactaws* (New York: Penguin, 1988), 30, 47; Richard T. Hughes and C. Leonard Allen, *Illusions of Innocence: Protestant Primi-*

tivism in America, 1630–1875 (Chicago: University of Chicago Press, 1988), xiii–xiv; also see Winton U. Solberg, "Primitivism in the American Enlightenment," in *The American Quest for the Primitive Church*, ed. Richard T. Hughes (Urbana: University of Illinois Press, 1988), 53–55.

28. John R. Knott, *Imagining Wild America* (Ann Arbor: University of Michigan Press, 2002), 33, 37–38, 42–43; Ella M. Foshay, *John James Audubon* (New York: Harry N. Abrams, 1997), 110–11; John James Audubon, *Delineations of American Scenery and Character* (New York: G. A. Baker, 1926), 4; Alton A. Lindsey, *The Bicentennial of John James Audubon* (Bloomington: University of Indiana Press, 1985), 115–24.

29. Knott, *Imagining Wild America*, 46–49; Foshay, *John James Audubon*, 111; Lee A. Vedder, *John James Audubon and the Birds of America* (San Marino, CA: Huntington Library, 2006), 21; also Marie R. Audubon, *Audubon and His Journals*, with zoological and other notes by Elliott Coues, vol. 1 (New York: C. Scribner's Sons, 1897), 182–83; Foshay, *John James Audubon*, 110–11.

30. Thoreau, *Journal* (New York, AMS Press, 1968), VIII, 220–21; III, 132–37.

31. Henry David Thoreau, *Walden and Other Writings* (New York: Modern Library, 1937), 191.

32. Thoreau, *Writings*, V, *Excursions and Poems*, 212; David G. Payne, *Voices in the Wilderness: American Nature Writing and Environmental Politics* (Hanover: University Press of New England, 1996), 36–37, 42–43; Thoreau, *Writings*, III, *The Maine Woods*, 133–35.

33. Thoreau, *Writings*, X, *Journal*, 23; *Writings*, XII, *Journal*, 387; Thoreau, "Huckleberries," in *The Natural History Essays* (Salt Lake City, UT: Peregrine Smith Books, 1980), 254, 259–60; Lawrence Buell, *The Environmental Imagination: Thoreau, Nature Writing, and the Formation of American Culture* (Cambridge: Belknap Press of Harvard University Press, 1995), 212–13.

34. Thoreau, *Writings, Journal*, II, 461–62.

35. Thoreau, *Journal*, XII, 152–55, April 21 and 22, 1859.

36. Laura Dassow, *Seeing New Worlds: Henry David Thoreau and Nineteenth-Century Natural Science* (Madison: University of Wisconsin Press, 1995), 4; David R. Foster, *Thoreau's Country: Journey through a Transformed Landscape* (Cambridge, MA: Harvard University Press, 1999), 13; David G. Payne, *Voices in the Wilderness: American Nature Writing and Environmental Politics* (Hanover, NH: University Press of New England, 1996), 36–37.

37. Thoreau, *Writings, Journal*, II, 461–62; XIV, 161, 262–63, 268, 293; also Donald Worster, *Nature's Economy: The Roots of Ecology* (Garden City: Anchor Books, 1979), 74; Henry David Thoreau, *Faith in a Seed: The Dispersion of Seeds and Other Late Natural History Writings*, ed. Bradley P. Dean (Washington, DC: Island Press, 1993), 134.

38. Daniel B. Botkin, *No Man's Garden: Thoreau and a New Vision for Civilization and Nature* (Washington, DC: Island Press, 2001), 120–23.

39. William R. Jordan III, *The Sunflower Forest: Ecological Restoration and the New Communion with Nature* (Berkeley: University of California Press, 2003), 153–59. Also W. R. Jordan III, "Renewal and Imagination: Thoreau's Thought and the Restoration of Walden Pond," in *Thoreau's World and Ours: A Natural Legacy*, ed. Edmund A. Schofield and Robert C. Baron (Golden, CO: North American Press, 1993). Literary critic Lawrence Buell discusses a "literary place creator" that resonates with the practice of restoration in a significant way in his *The Environmental Imagination*. For an extensive treatment of Thoreau's idea of redemption through imagination see

Frederick Garber, *Thoreau's Redemptive Imagination* (New York: New York University Press, 1977).

40. David Lowenthal, *George Perkins Marsh: Prophet of Conservation* (Seattle: University of Washington Press, 2003), 17–19.

41. Hall, *Earth Repair* (Charlottesville: University of Virginia Press, 2005), 23.

42. Ibid., 19ff.

43. Michael P. Cohen, *The Pathless Way: John Muir and the American Wilderness* (Madison: University of Wisconsin Press, 1981), 19–20, 191; John Muir Papers, 1858–1957 (Microfilm Edition, University of the Pacific), Series II, Journals and Sketchbooks, Reel #23, 150, 154. On Muir, also see Stephen Fox, *The American Conservation Movement: John Muir and His Legacy* (Boston: Little, Brown, 1981).

44. Bruce Peneck, "Saving Wilderness for Sacramental Use: John Muir," in *Conservation Reconsidered: Nature, Virtue, and American Liberal Democracy,* ed. Charles T. Rubin (Lanham, MD: Rowan and Littlefield, 2002), 21.

45. Peneck, "Saving Wilderness," 22.

46. Bill Devall and George Sessions, "Development of Natural Resources," *Environmental Ethics* 6 (Winter 1984): 306, 316; see also Ray Raphael, *Tree Talk: The People, Politics, and Economics of Timber* (Washington, DC: Island Press, 1981).

47. Char Miller, *Gifford Pinchot and the Making of Modern Environmentalism* (Washington, DC: Island Press/Shearwater Books, 2001), 155.

Chapter 3: Preconditions

1. This notion has a long history. Coasting off Venezuela on his third voyage to the Americas in 1498, Christopher Columbus wrote, "There are indications of this being the terrestrial paradise." He was sounding a note that would play an important role in the ensuing mythology. Christopher Columbus, *Four Voyages to the New World*, trans. R. H. Major (Gloucester, MA: Corinth, 1978), 135.

2. Ed Folsom, "Walt Whitman's Paradise Regained," in *Recovering the Prairie*, ed. Robert F. Sayre (Madison: University of Wisconsin Press, 1999), 47–60.

3. Willa Cather, *O Pioneers!* (Cambridge, MA: Houghton Mifflin, 1962),75.

4. Leo Marx, personal communication with William R. Jordan III, April 1992.

5. David Lowenthal, *The Past Is a Foreign Country* (Cambridge: Cambridge University Press, 1985), 389.

6. Joni L. Kinsey, Rebecca Roberts, and Robert F. Sayre, "Prairie Prospect: The Aesthetics of Plainness," in *Recovering the Prairie*, ed. Robert F. Sayre (Madison: University of Wisconsin Press, 1999), 14–45.

7. Roderick Nash, *Wilderness and the American Mind*, 3rd ed. (New Haven, CT: Yale University Press, 1982), 145.

8. Ibid., 43.

9. Bruce Wilshire, *The Primal Roots of American Philosophy: Pragmatism, Phenomenology and Native American Thought* (University Park: Pennsylvania State University Press, 2000).

10. Carolyn Merchant, *Reinventing Eden: The Fate of Nature in Western Culture* (New York: Routledge, 2003); also "Reinventing Eden: Western Culture as a Recovery Narrative," in *Uncommon Ground: Toward Reinventing Nature,* ed. William Cronon (New York: W.W. Norton, 1995), 132–59, esp. 155–56; also see David E. Nye, "Technology, Nature and American Origin Stories," *Environmental History* 8 (January 2003): 8–24. There is an extensive literature on the influence of the edenic ideal and

the theme of recovery in American thought. For further treatment see Anne Whiston Spirn, "Constructing Nature: The Legacy of Frederick Law Olmsted," and Candace Slater, "Amazonia as Edenic Narrative," in *Uncommon Ground*; also R. W. B. Lewis, *The American Adam: Innocence, Tragedy and Tradition in the Nineteenth Century* (Chicago: University of Chicago Press, 1955), and John Armstrong, *The Paradise Myth* (Oxford: Oxford University Press, 1969).

11. Leo Marx, *The Machine in the Garden: Technology and the Pastoral Ideal in America* (New York: Oxford University Press, 1964).

12. Donald Worster, *The Wealth of Nature: Environmental History and the Ecological Imagination* (Oxford: Oxford University Press, 1993), 184–202.

13. Ibid. In *A Passion for Nature: The Life of John Muir* (Oxford: Oxford University Press, 2008), 8–10, Worster points out that in his *Democracy in America*, Alexis de Tocqueville noted the tendency of American democracy to extend the circle of value and entitlement beyond humans and to replace the doctrines of Christianity with a pantheism that accorded value to nonhuman creatures.

14. David Lowenthal, "Reflections on Humpty-Dumpty Ecology," in *Restoration and History: The Search for a Useable Environmental Past*, ed. Marcus Hall (New York: Routledge, 2010), 13–24.

15. David Lowenthal, *The Past Is a Foreign Country* (Cambridge: Cambridge University Press, 1985).

16. See for example Samuel P. Hays, *Conservation and the Gospel of Efficiency: The Progressive Conservation Movement, 1890–1920* (New York: Atheneum, 1975).

17. Quoted in Al Runte, *National Parks: The American Experience* (Lincoln: University of Nebraska Press, 1987), 204.

18. Richard West Sellars, *Preserving Nature in the National Parks* (New Haven, CT: Yale University Press, 1997), 4.

19. Marcus Hall, *Earth Repair: A Transatlantic History of Environmental Restoration* (Charlottesville: University of Virginia Press, 2005), 106ff.

20. Élisée Reclus, *The Ocean Atmosphere and Life: A Descriptive History of the Phenomena of the Life of the Globe*, 1872, trans. and ed. A. H. Keane (London: J. S. Virtue, 1887).

21. Frederick Law Olmsted and Calvert Vaux, "General Plan for the Improvement of the Niagara Reservation. 1887." This document is available online at www.niagaraheritage.org/genplan.htm.

22. Richard Sellars makes this point in *Preserving Nature in the National Parks: A History* (New Haven, CT: Yale University Press, 1997), 15–16.

23. Reid Beveridge, personal communication, February 3, 2009.

24. Frank A. Waugh, *The Natural Style in Landscape Gardening* (Boston: R.A. Badger, 1917), 21–22, 24; also Waugh, "American Ideals in Landscape Architecture," *Landscape Architecture* 15 (April 1925): 151–54; Robert E. Grese, "Historical Perspectives on Designing with Nature," in *Restoration '89: The New Management Challenge*, First Annual Meeting of the Society for Ecological Restoration (Oakland, CA: Society for Ecological Restoration/University of Wisconsin Press, 1989), 45.

25. Robert E. Grese, *Jens Jensen: Maker of Natural Parks and Gardens* (Baltimore: Johns Hopkins University Press, 1992), 7–9; Leonard K. Eaton, *Landscape Artist in America: The Life and Work of Jens Jensen* (Chicago: University of Chicago Press, 1964), 17.

26. Robert E. Grese, "Jens Jensen: The Landscape Artist as Conservationist," in *Midwestern Landscape Architecture*, ed. William H. Tishler (Urbana: University of

Illinois Press, 2000), 117, 131; Jens Jensen, "Landscape Art: An Inspiration from the Western Plains," *Sketch Book* 6 (1906): 22.

27. Grese, "Historical Perspectives," 43–44; Grese, "Jens Jensen: The Landscape Artist as Conservationist," 127; Victor M. Cassidy, *Henry Chandler Cowles: Pioneer Ecologist* (Chicago: Kedzie Sigel Press, 2007), 63–68.

28. Grese, "Jens Jensen: The Landscape Artist as Conservationist," 120–21, 127–28; Grese, "Historical Perspectives," 42–43.

29. Julia Sniderman Bachrach, "Ossian Cole Simonds: Conservation Ethic in the Prairie Style," in *Midwestern Landscape Architecture*, ed. William H. Tishler (Urbana: University of Illinois Press, 2000), 80, 87–88; Grese, "Historical Perspectives," 40.

30. Grese, "Historical Perspectives," 40–41.

31. Wilhelm Miller, *The Prairie Spirit in Landscape Gardening* (Amherst: University of Massachusetts Press, 2002), 10–11, 15–16. Miller, along with contemporaries such as Jens Jensen, also promoted the adoption of naturalizing motifs for areas such as home landscapes and rights-of-way. See Miller's "The Illinois Way of Beautifying the Farm" (Urbana: Agricultural Experiment Station, University of Illinois, Circular 170, 1914).

32. Eugene Cittadino, "Ecology and American Social Thought," in *Religion and the New Ecology: Environmental Responsibility in a World of Flux*, ed. David M. Lodge and Christopher Hamlin (Notre Dame, IN: Notre Dame University Press, 2006), 73–115.

33. Eugene Cittadino, personal communication with William R. Jordan III, July 24, 2009.

34. Mark Stoll, "Creating Ecology," in *Religion and the New Ecology: Environmental Responsibility in a World of Flux*, ed. David M. Lodge and Christopher Hamlin (Notre Dame, IN: Notre Dame University Press, 2006), 56.

35. Ibid.

36. Stephen A. Forbes, "On Some Interactions of Organisms," *Illinois Laboratory of Natural History, Bulletin* 1: 3–17, quoted in Cittadino, "Ecology and American Social Thought," 79.

37. Mircea Eliade, *The Myth of the Eternal Return, or Cosmos and History* (Princeton, NJ: Princeton University Press, 1954).

38. For a discussion of this idea and an interpretation of restoration as an attempt to reconcile the ideas of "improvers" and romantic regard for wilderness, see Devin DeWayne Corbin, *The Work of Belonging: Agricultural Improvement, Romantic Wilderness, and the Rise of Restorationism in United States Environmental Literature* (Ph.D. dissertation, University of Minnesota, 2005). Also see Nash, *Wilderness and the American Mind*, 25.

39. Richard West Sellars discusses the Organic Act and other aspects of the early development of the national parks in *Preserving Nature in the National Parks: A History* (New Haven, CT: Yale University Press, 1997).

40. Ibid., 22.

41. M. D. Spence, *Dispossessing the Wilderness: Indian Removal and the Making of the National Parks* (New York: Oxford University Press, 1999).

42. Kat Anderson, *Tending the Wilderness: Native American Knowledge and the Management of California's Natural Resources* (Berkeley: University of California Press, 2005); Charles Mann, *New Revelations of the Americas before Columbus* (New York: Knopf, 2005).

43. Sellars, *Preserving Nature*, 23, 82ff, 124.

44. Hall, *Earth Repair*, 142, 196.

45. Ben Thompson to Arno B. Cammerer, February 23, 1934. George M. Wright files, MVZ-UC. The statement later appeared in *Fauna No. 2* (Washington, DC: Government Printing Office, 1935), 123–24; quoted in Sellars, *Preserving Nature*, 92.

46. U.S. Department of the Interior, National Park Service, *Fauna of the National Parks of the United States: A Preliminary Survey (Fauna No. 1)*, by George M. Wright, Joseph S. Dixon, and Ben H. Thompson (Washington, DC: Government Printing Office, 1933), 21; quoted in Alfred Runte, *Yosemite: The Embattled Wilderness* (Lincoln: University of Nebraska Press, 1990), 162.

47. Ibid., 10.

48. Ibid., 21.

49. Ibid., 37.

50. National Park Service, "Policy on Predators and Notes on Predators," 1939, various pagination, Central Classified File, RG79; quoted in Sellars, *Preserving Nature*, 98.

51. Wright et al., *Fauna No. 1*, 69–70; Sellars, *Preserving Nature*, 97, 139.

52. George M. Wright, Joseph S. Dixon, and Ben H. Thompson, *Fauna of the National Parks of the United States, Fauna Series No. 2* (Washington, DC: U.S. Government Printing Office, 1935), 3.

53. See for example, Barbara Novak, *Nature and Culture: American Landscape and Painting, 1825–1875* (Oxford: Oxford University Press, 1980), 30ff.

54. Charles C. Adams, "Ecological Conditions in National Forests and in National Parks," *Scientific Monthly* 20 (June 1925): 570; quoted in Sellars, *Preserving Nature*, 90.

Chapter 4: Invention

1. Jacques Barzun, *Science: The Glorious Entertainment* (New York: Harper & Row, 1964).

2. F. V. Coville and D. T. MacDougal, *The Desert Botanical Laboratory of the Carnegie Institution* (Washington, DC: Carnegie Institution, 1903), No. 6.

3. Howard S. Reed, "Volney Morgan Spalding," *The Plant World* 22 (1919): 14–18.

4. V. M. Spalding to D. T. MacDougal, May 28, 1905, Arizona Historical Society, Tucson.

5. D. T. MacDougal, "Botanical Research: The Desert Botanical Laboratory Tucson, Arizona," in Carnegie Institution of Washington Year Book No. 5, Washington, DC, 1906, 119–134.

6. Forrest Shreve, "Changes in Desert Vegetation," *Ecology* 10 (1929): 364–73. Reprinted from *Progressive Arizona and the Great Southwest*, 3–12, ca. 1930.

7. Hall, *Earth Repair*, 103–19, 121–33 for Sampson and 113–16, 119–24 for Ellison.

8. Michael Rosenzweig, *Win–Win Ecology: How the Earth's Species Can Survive in the Midst of Human Enterprise* (New York: Oxford University Press, 2002).

9. Michael Rosenzweig, personal communication with William R. Jordan III, August 2008.

10. Edith A. Roberts, "The Development of an Out-of-Door Botanical Laboratory for Experimental Biology," *Ecology* 14:2 (April 1933): 163.

11. Emily Bloom Griswold, "The Origin and Development of Ecogeographic Displays in North American Botanic Gardens" (master's thesis, University of Washington, 2002).

12. Meg Ronsheim, personal communication with William R. Jordan III, July 21, 2009.

13. Conrad Liegel, "Arboretum Citations," Winter 1977, typescript of publications related to the Cowling Arboretum, Carleton College Archives, p. 1. Thanks to Dave Egan for drawing our attention to this project.

14. "The Carleton Arboretum: A Century of Constant Change," *The Carleton Voice* 43:4 (Summer 1978): 4.

15. Leal A. Headley and Merrill E. Jarchow, *Carleton: The First Century* (Northfield, MN: Carleton College, 1966), 35–36. The information about the scale of the project was provided by Cowling Arboretum Director Nancy C. Braker.

16. Memo, Item 8, General Study, Item 8, p. 5 in "Progress Report on the Holden Arboretum, October 6, 1932"; photocopy in Box 41, Holden Arboretum archives.

17. Arthur B. Williams, "The Cleveland Museum of Natural History, Holden Arboretum, Report of Progress for 1935" (April 1936), 4. Ibid.

18. Letter from Harold Madison to Mrs. Benjamin Patterson Bole, April 11, 1935. Attachment to Minutes of Board Meeting, April 11, 1935. Ibid.

19. Ibid.

20. Williams, "The Cleveland Museum of Natural History Report for 1935," 4.

21. Letter from E. D. Merrill to Warren Corning, December 15, 1938. Attachment to Minutes of Board Meeting, December 28, 1938.

22. Williams, "Cleveland Museum of Natural History Report of Progress for 1935," 17.

23. Aldo Leopold, "Natural History," in *A Sand County Almanac with Essays on Conservation from Round River* (New York: Ballantine, 1966), 202–10.

24. Ambrose Crawford, a one-page typed summary of the project dated June 1979, Alstonville Plateau Historical Society, Alstonville, NSW.

25. Horace Webber, *The Greening of the Hill: Revegetation around Broken Hill in the 1930s* (Victoria, Australia: Hyland House, 1992). We are also drawing from Dr. McDonald's notes of her investigation and interview with Dorothy Crawford during the summer of 2008.

26. Nancy D. Sachse, *A Thousand Ages: The University of Wisconsin Arboretum* (Madison: University of Wisconsin Arboretum, 1974), 16ff.

27. Aldo Leopold, "What Is the University of Wisconsin Arboretum, Wild Life Refuge, and Forest Experiment Preserve?" Address at the Dedication of the Arboretum, June 17, 1934, in William R. Jordan III, ed., *Our First 50 Years: The University of Wisconsin Arboretum* (Madison: University of Wisconsin, 1984).

28. Sachse, *A Thousand Ages*, 45ff.

29. Ibid., pp. 32–33.

30. Leopold, "What Is the University of Wisconsin Arboretum," 2–5.

31. Curt Meine, *Aldo Leopold: His Life's Work* (Madison: University of Wisconsin Press, 1988), 375.

32. Frank Cook, personal communication, July 21, 2010.

33. Sachse, *A Thousand Ages*, 49; also Thomas J. Blewett and Grant Cottam, "History of the University of Wisconsin Arboretum Prairies," *Transactions of the Wisconsin Academy of Sciences, Arts and Letters* 72 (1984): 130–44.

34. R. D. Linton, "Arboretum Trees and Shrubs Are Longenecker's Children," *The (Madison, WI) Capital Times*, October 22, 1933, cited in Emily Bloom Griswold, "The Origin and Development of Ecogeographic Displays," 81.

35. Philip J. Pauly, *Fruits and Plains: The Horticultural Transformation of America* (Cambridge, MA: Harvard University Press, 2007), 190–94.

36. Werner O. Nagel to Paul V. Brown, October 10, 1935. UW Archives, 38/3/2, Box 2. (Thanks to Frank Court for referring us to this document.)

37. Theodore M. Sperry, "Prairie Restoration on the University of Wisconsin Arboretum. Spring, 1939." University of Wisconsin archives. University of Wisconsin Arboretum. Correspondence folder, 38/7/2, Box 1.

38. For a detailed account of resistance to the use of fire as a land management tool by Americans during this period, see Stephen J. Pyne, *Fire in America: A Cultural History of Wildland and Rural Fire* (Seattle: University of Washington Press, 1997).

39. Popular and agency resistance to fire and to burning as a land management tool was an important element in controversies over conservation throughout the twentieth century. For overviews, see Pyne, *Fire in America*, 197, 260–94; Christopher J. Huggard and Arthur R. Gomez, eds., *Forests under Fire: A Century of Ecosystem Mismanagement in the Southwest* (Tucson: University of Arizona Press, 2001), xv, xxvii–xxx, 183. See also Paul Hirt, *Conspiracy of Optimism: Management of the National Forests Since World War II* (Lincoln: University of Nebraska Press, 1994).

40. J. T. Curtis and M. L. Partch, "Effect of Fire on the Competition between Bluegrass and Certain Prairie Plants," *The American Midland Naturalist* 39, no. 2 (1948): 237–43.

41. Aldo Leopold, letter to Paul Brown, February 10, 1940. UW Archives, Arboretum, Administrative personnel, 9/25/10-5, Box 1.

42. This interpretation has become a commonplace in popular accounts of aspects of conservation history. See Kenneth Brower, "Leopold's Gift," *Sierra* (January/February 2001), 30–39, 109–10. For a recent example see Rocky Barker, *Scorched Earth: How the Fires of Yellowstone Changed America* (Washington, DC: Island Press, 2005), 140–42.

43. Eric Higgs, *Nature by Design: People, Natural Process and Ecological Restoration* (Cambridge, MA: MIT Press, 2003), 83–84.

Chapter 5: Neglect

1. Ron Patkus, Vassar College, personal communication with William R. Jordan III, October 9, 2008. Patkus notes that Botany Department records indicate that Professor Roberts retired in 1948 and believes that an obituary indicating that the year was 1950 is erroneous.

2. Dorothy Wurman, "The Dutchess County Botanical Garden: A First in the United States." Initial draft, January 24, 2002. Link to this document provided by Keri Van Camp: /kevancamp/Eco Gardens/Ecological Garden PDF.pdf. https://vspace .vassar.edu/xythoswfs/webui/_xy-1581469_1-t_ylmrO23V.

3. J. Baird Callicott, "Whither Conservation Ethics?," in *Beyond the Land Ethic: More Essays in Environmental Philosophy* (Albany: State University of New York Press, 1999), 321–31.

4. Aldo Leopold, "The Land Ethic," in *A Sand County Almanac with Essays from Round River* (New York: Ballantine, 1966), 245.

5. Ibid., 244.

6. Aldo Leopold, *Ecology and Economics*, Undated manuscript, University of Wisconsin Archives, 9/25/10-6, Box 17, book 81-198.

7. Ibid., 203.

8. Leopold, "What Is the University of Wisconsin Arboretum, Wild Life Refuge, and Forest Experiment Preserve?," in *Our First 50 Years: The University of Wisconsin Arboretum, 1934–1984*, ed. William R. Jordan III (Madison: University of Wisconsin Arboretum, 1984), 2–5.

9. Leopold, *A Sand County Almanac*, xviii.

10. Callicott, "Whither Conservation Ethics?," 328.

11. Bryan Norton, *Toward Unity among Environmentalists* (New York: Oxford University Press, 1991), 57.

12. Curt Meine, "Leopold's Fine Line," in *Correction Lines: Essays on Land, Leopold and Conservation* (Washington, DC: Island Press, 2004), 80–116.

13. J. Baird Callicott, "From the Balance of Nature to the Flux of Nature: The Land Ethic in a Time of Change," in *Aldo Leopold: An Ecological Conscience*, ed. Richard L. Knight (Washington, DC: Island Press, 2002), 91–105.

14. Eugene Hargrove, *Foundations of Environmental Ethics* (Englewood Cliffs, NJ: Prentice Hall, 1989), 137–64. Susan Flader deals with the development of Leopold's thinking about management of ecological systems, especially with reference to deer, in *Thinking Like a Mountain: Aldo Leopold and the Evolution of an Ecological Attitude toward Deer, Wolves and Forests* (Columbia: University of Missouri Press, 1974).

15. Hargrove, *Foundations of Environmental Ethics*, 151.

16. Ibid., 153. See also J. Baird Callicott and Eugene C. Hargrove, "Leopold's Means and Ends in Wildlife Management: A Brief Commentary," *Environmental Ethics* 12, no. 4 (Winter 1990): 333–37.

17. Donald Worster, *Nature's Economy: The Roots of Ecology* (Garden City, NY: Anchor Books, 1979), 289–90. Worster is not alone in this characterization of Leopold's mature thinking. See Harold Fromm, "Aldo Leopold: Aesthetic Anthropocentrist," *Isle* (Spring 1993), reprinted in *The Isle Reader* (Athens: University of Georgia Press, 2003). See also J. Baird Callicott and Eugene Hargrove in "Leopold's Means and Ends in Wildlife Management," especially their discussion (336) of arguments of Scott Lehmann and Richard C. Watson that Leopold's mature philosophy was fundamentally anthropocentric.

18. Callicott, "The Arboretum and the University: The Speech and the Essay," *Transactions of the Wisconsin Academy of Sciences, Arts, and Letters* 87 (1999): 13.

19. By characterizing the kind of activity Leopold recounts in *A Sand County Almanac* as "play" we do not mean to downplay its importance relative to work. Quite the contrary: What we have in mind is the kind of experience philosopher Josef Pieper characterizes as "leisure," which he sees as prior to work in importance and as essential to the contemplation of the world that he argues is the "basis of culture." Josef Pieper, *Leisure the Basis of Culture* (San Francisco: Ignatius Press, 2009). It is also worth noting here the role hunters and anglers have played in conservation. Historian John Reiger argues that "those who hunted and fished for pleasure" (21)—people seriously engaged in one of the classic forms of play—were not just the beneficiaries but "the real spearhead" of this movement. See John F. Reiger, *American Sportsmen and the Origins of Conservation* (Norman: University of Oklahoma Press, 1986).

20. Leopold, *A Sand County Almanac*, xviii.

21. Ibid., xviii.

22. Ibid., 52.

23. Bill Jordan recalls discussing this point with his father, a forester, around the dinner table in connection with the work at the UW–Madison Arboretum in the early 1960s.

24. Worster, *Nature's Economy*, 289–99.

25. David Lowenthal, *The Past Is a Foreign Country* (Cambridge: Cambridge University Press, 1985), 391.

26. Arthur Herman, *The Idea of Decline in Western History* (New York: The Free Press, 1997).

27. Terrell Dixon, "Toxicity and Restoration: What American Literature since Rachel Carson Tells Us." Paper presented at the Tenth Annual Conference of the Society for Ecological Restoration, Austin, TX, 1998; Conference Abstracts, 74.

28. Aldo Leopold, "Wilderness as a Land Laboratory," in *The River of the Mother of God*, ed. Susan L. Flader and J. Baird Callicott (Madison: University of Wisconsin Press, 1991), 287–89.

29. Peter Friederici, "One Man, Fifteen Acres, Forty Years," in *Nature's Restoration: People and Places on the Front Lines of Conservation* (Washington, DC: Island Press, 2006), 1–33.

30. Richard West Sellars, *Preserving Nature in the National Parks* (New Haven, CT: Yale University Press, 1997), 4. In *Playing God in Yellowstone: The Destruction of America's First National Park* (Boston: Atlantic Monthly Press, 1986), Alston Chase takes the National Park Service to task for its management policies, which he critiques as reflecting a vague pantheism emphasizing the experience of oneness with the world. In a more recent book, *In a Dark Wood: The Fight over Forests and the Myths of Nature* (New Brunswick, NJ: Transaction Press, 2001), Chase returns to this theme and argues that all value is a cultural construct and that for this reason the idea of altruistic land management is untenable and all management is self-interested stewardship. Interestingly, Chase characterizes the early work at the UW–Madison Arboretum as consistent with this idea and criticizes the 1963 Leopold Report as betraying it (pp. 112ff).

31. This discussion of the Leopold Report is based on Sellars's treatment in *Preserving Nature*, 195ff. The report was published as A. Starker Leopold et al., "Wildlife Management in the National Parks," in *Transactions of the Twenty-Eighth North American Wildlife and Natural Resources Conference*, ed. James B. Trerethen (Washington, DC: Wildlife Management Institute, 1963). Historian Stephen Pyne's account of the Park Service's ambivalent response to proposals for the "oxymoronic 'prescribed natural fire'" (252) also underscores both the conceptually problematic nature of this essential restorative technology and the difficulty an agency such as the NPS had in responding to it effectively. See Pyne's *Vestal Fire: An Environmental History, Told through Fire, of Europe and Europe's Encounter with the World* (Seattle: University of Washington Press, 1997), 442–52.

32. Robert M. Linn, "The Natural Resources Committee: A Functional Concept"; Linn, "The Science Program in the National Park Service"; Acting Assistant Director to S. Herbert Evison, September 13, 1966. NPS History Collection, Harpers Ferry, WV; Sumner, "A History of the Office of Natural Science Studies," 1. Cited in Sellars, *Preserving Nature*, 223.

33. Sellars, *Preserving Nature*, 214; Lowell Sumner, "A History of the Office of Natural Science Studies," in *Proceedings of the Meeting of Research Scientists and Management Biologists of the National Park Service*, Horace M. Albright Training Center, April 6–8, 1968, typescript, 4, Dennis; quoted in Sellars, 201.

34. Victor E. Shelford, ed., *Naturalist's Guide to the Americas* (Baltimore: Williams & Wilkins, 1926).

35. The edited volume that resulted from this conference also reflects the oil-and-water relationship between restoration and preservation at this time. The index lists only two references to restoration in chapters outside the section on restoration. E. O. Wilson and Francis M. Peter, eds., *Biodiversity* (Washington, DC: National Academy Press, 1988).

36. Eric Higgs, *Nature by Design: People, Natural Processes, and Ecological Restoration* (Cambridge, MA: MIT Press, 2003), 217.

37. Michael Fischer, personal communication with William R. Jordan III, September 1, 2010.

38. J. B. Zedler, G. D. Williams, and J. Desmond, "Wetland Mitigation: Can Fishes Distinguish between Natural and Constructed Wetlands?," *Fisheries* 22 (1998): 26–28; J. B. Zedler, "Replacing Endangered Species Habitat: The Acid Test of Wetland Ecology," in *Conservation Biology for the Coming Age*, ed. P. L. Fiedler and P. M. Kareiva (New York: Chapman and Hall, 1998), 364–79. Also see the exchange between Zedler and John Rieger in *Restoration & Management Notes* (1991): 65–68.

39. A. Dwight Baldwin, Judith DeLuce, and Carl Pletsch, eds., *Beyond Preservation: Restoring and Inventing Landscapes* (Minneapolis: University of Minneapolis Press, 1994). For discussion of the restoration–management distinction, see Orie L. Loucks's essay, 127–35 and W. R. Jordan's response, 244–46. For comments especially critical of restoration, see the essays by Jack Temple Kirby, Constance Pierce, and G. Stanley Kane.

40. Orie L. Loucks, "Art and Insight in Remnant Native Ecosystems," in Baldwin et al., *Beyond Preservation*, 127–35.

41. Jared Diamond, "Is It Necessary to Shoot Deer to Save Nature?" *Natural History Magazine* (August 1992): 2–8.

42. Kevin McSweeney, Arboretum Director, personal communication with William R. Jordan III, November 2008. See also Madeline Fisher, "For All the Right Seasons," *On Wisconsin Magazine* (Fall 2009); onwisconsin.uwalumni.com/features/for-all-the-right-seasons/.

43. Grant Cottam and Roger C. Anderson, "Proposal for an Environmental Research Facility at the University of Wisconsin Arboretum"; undated, requests funds for the period December 1, 1971–June 1, 1973. Document in arboretum files.

Chapter 6: Realization I: Stepping-Stones

1. See Eugene Cittadino, "Ecology and American Social Thought," in *Religion and the New Ecology: Environmental Responsibility in a World in Flux*, ed. David M. Lodge and Christopher Hamlin (Notre Dame, IN: Notre Dame University Press, 2006), 73–115.

2. See Elmer B. Hadley and Barbara J. Kiekhefer, "Productivity of Two Prairie Grasses in Relation to Fire Frequency," *Ecology* 44, no. 2 (1969): 389–55. This account of the early work at Trelease Prairie is based on interviews with Mr. Buck, March 26 and September 22, 2009.

3. Paul Shepard, "Green Oaks and Knox College: A Look at the Future. A Proposal for the Development of Green Oaks." Appendix VI, March 1956. Unpublished typescript. Knox College archives, Green Oaks, Box 2.

4. Paul Shepard, "Green Oaks: A Memoir." Undated manuscript, p. 9. Green Oaks archive. Knox College Library, Box 1.

5. Paul Shepard interview by William R. Jordan III, November 9, 1988, audio recording AC 126 and transcript in UW–Madison Arboretum oral history archives. The material quoted is on p. 4 of the transcript. Bruce Allison, a professor of biology at Knox College, notes that the cabin was built but burned down sometime in the early 1980s.

6. Ibid., 8.

7. Shepard, "Green Oaks and Knox College," 1.

8. Shepard, "Green Oaks: A Memoir," 32.

9. Shepard, "Green Oaks and Knox College," 3–4.

10. Shepard, "Green Oaks: A Memoir," 29–30.

11. Shepard, "Green Oaks and Knox College," 6.

12. Peter Schramm interview by William R. Jordan III, Knox College, May 22, 2007. Information about the North American Prairie Conferences is available online at www.unk.edu/nss/biology.aspx?id=14129.

13. Paul Shepard, *The Tender Carnivore and the Sacred Game* (New York: Charles Scribner's Sons, 1973), 269.

14. Ibid., 260ff.

15. John Cairns Jr. interview by William R. Jordan III, October 17, 2003. Also see John Cairns Jr., "Restoration and Management: An Ecologist's Perspective," *Restoration & Management Notes* 1, no. 1 (1980): 6–8.

16. Aldo Leopold, "Wilderness," in *A Sand County Almanac with Essays from Round River* (New York: Ballantine, 1966), 264–79. The quote is on p. 278.

17. John J. Berger, *Restoring the Earth: How Americans Are Working to Renew Our Damaged Environment* (New York: Doubleday, 1987), 116–17. Berger's book was originally published in 1985.

18. William K. Stevens, *Miracle under the Oaks: The Revival of Nature in America* (New York: Pocket Books, 1995), 107ff; Ray Schulenberg, "Prairie in a Post-Prairie Era," *The Morton Arboretum Quarterly* 3, no. 2 (Summer 1967): 17–27.

19. Stevens, *Miracle under the Oaks*, 109ff; Harold L. Nelson, "Prairie Restoration in the Chicago Area," *Restoration & Management Notes* 5, no. 2 (Winter 1987): 60–67. See also Fermilab's website at history.fnal.gov/prairie.html#ecological_front.

Chapter 7: Realization II: Taking Hold

1. Bill McKibben provides an overview of the recovery of the forests of the Northeast since the mid-nineteenth century in the first chapter, "Home," of his book *Hope, Human and Wild: True Stories of Living Lightly on the Earth* (Boston: Little, Brown, 1995).

2. Leslie Sauer, interview with William R. Jordan III, September 25, 2008. The preceding material concerning Leslie Sauer and Andropogon is based on this interview. Eric Higgs expresses a similar skepticism about the effects of legislation and regulation on restoration in *Nature by Design: People, Natural Processes, and Ecological Restoration* (Cambridge, MA: MIT Press, 2003), 207.

3. In contrast with the midwestern prairies, where early efforts at restoration were carried out by amateurs or by ecologists rather than professional managers and had little relation to the in some ways relevant practice of range management.

4. C. B. Craft, J. Bertram, and S. Broome, "Coastal Zone Restoration," in *Ecological Engineering*, ed. S. E. Jorgenson and B. D. Fath (Oxford: Elsevier B.V., 2008), 637–44.

5. Cynthia Zaitzevsky, *Frederick Law Olmsted and the Boston Park System* (Boston: Belknap, 1982); Dave Egan, "Historic Initiatives in Ecological Restoration," *Restoration & Management Notes* 8, no. 2 (1990): 83–90.

6. André F. Clewell and James Aronson, *Ecological Restoration: Principles, Values, and Structure of an Emerging Profession* (Covelo, CA: Island Press, 2007), 70–74.

7. John Berger, *Ecological Restoration in the San Francisco Bay Area: A Descriptive Directory and Sourcebook* (Berkeley: Restoring the Earth, 1990), 142; H. T. Harvey, "Some Ecological Aspects of San Francisco Bay," prepared for San Francisco Bay Conservation and Development Commission, October 1966; H. T. Harvey, P. Williams, and J. Haltiner, "Guidelines for Enhancement and Restoration of Diked Historic Baylands," prepared for San Francisco Bay Conservation and Development Commission, April 1982; R. J. Hartesveldt and H. T. Harvey, "The Fire Ecology of Sequoia Regeneration," *Tall Timbers Fire Ecology Conference Proceedings* 7 (1967): 65–77.

8. Prof. Arnold Van der Valk, interview with William R. Jordan III, July 2, 2009.

9. S. M. Galatowitsch and A. G. van der Valk, "The Vegetation of Restored and Natural Prairie Wetlands," *Ecological Applications* 6 (1996): 102–12; A. G. van der Falk, "Succession Theory and Restoration of Wetland Vegetation," in *Wetlands for the Future*, ed. A. J. McComb and J. A. Davis (Adelaide, Australia: Gleneagles, 1998), 657–67.

10. Rich Reiner and Tom Griggs, "Nature Conservancy Undertakes Riparian Restoration Projects in California," *Restoration & Management Notes* 7, no. 1 (1989): 3–8.

11. See *Compensating for Wetland Losses under the Clean Water Act* (Washington, DC: National Academy Press, 2001), especially the Executive Summary, pp. 1–10.

12. Robin Freeman interviews with William R. Jordan III, May 15 and 29, 2009.

13. *San Francisco Estuary Project: Comprehensive Conservation and Management Plan* (San Francisco: San Francisco Bay Regional Water Quality Control Board, 1993).

14. Carole Schemmerling, interview with William R. Jordan III, July 7, 2009.

15. See Finn Wilcox and Jerry Gorsline, eds., *Working the Woods, Working the Sea: An Anthology of Northwest Writing* (Port Townsend, WA: Empty Bowl, 2008).

16. See Timothy Egan's *The Good Rain: Across Time and Terrain in the Pacific Northwest* (New York: Knopf, 1990).

17. Billy Frank, interview with William R. Jordan III, September 8, 2010.

18. Dean Apostol and Marcia Sinclair, *Restoring the Pacific Northwest: The Art and Science of Ecological Restoration in Cascadia* (Washington, DC: Island Press, 2006).

19. Dean Apostol, interview with William R. Jordan III, September 23, 2009.

20. Jerry Gorsline, "Treeplanter's Journal," in *Working the Woods, Working the Sea: An Anthology of Northwest Writing*, ed. Finn Wilcox and Jerry Gorsline (Port Townsend, WA: Empty Bowl, 2008), 78–84.

21. "Forest Ecosystem Management: An Ecological, Economic and Social Assessment." Report of the Forest Ecosystem Management Team (Washington, DC: The Service, 1993).

22. Dean Apostol, interview with William R. Jordan III.

23. Apostol and Sinclair, *Restoring the Pacific Northwest*, 16.

24. Steve Moddemeyer, interview with William R. Jordan III, May 27, 2009.

25. This section is based in part on conversations with Martinez in September and October 2010.

26. Kat Anderson, *Tending the Wild: Native American Knowledge and the Management of California's Natural Resources* (Berkeley: University of California Press, 2005), 1.

27. Thomas Davis, *Sustaining the Forest, the People and the Spirit* (Albany: SUNY Press, 2000).

28. Donald Waller, personal communication to William R. Jordan III, November 1, 2010. Also see Donald M. Waller and Thomas P. Rooney, *The Vanishing Present: Wisconsin's Changing Lands, Water and Wildlife* (Chicago: University of Chicago Press, 2008).

29. See extensive explorations of this idea by Gary Nabhan in *Cultures of Habitat* (Washington, DC: Counterpoint, 1998) and by Anderson in *Tending the Wild*. For discussion of the implications of this work for management of national parks, see the December 2003 issue of *Ecological Restoration*.

30. David Tomblin, *Managing Boundaries, Healing the Homeland: Ecological Restoration and the Revitalization of the White Mountain Apache Tribe* (Ph.D. dissertation, Virginia Technical Institute, 2009); see especially chapter 3.

31. Rafael Sagarin and Anibal Pauchard, "Observational Approaches in Ecology Open New Ground in a Changing World," *Frontiers in Ecology and the Environment* 8, no. 7 (2008): 379–86; Jeffrey E. Herrick et al., "National Ecosystem Assessment Supported by Scientific and Local Knowledge," *Frontiers in Ecology and the Environment* 8, no. 8 (2008): 403–8.

32. Jane George, "New Bowhead Numbers Show Inuit Are Right," *Nunatsiaq News*, March 14, 2008.

33. See M. D. Spence, *Dispossessing the Wilderness: Indian Removal and the Making of the National Parks* (New York: Oxford University Press, 1999).

34. Canada and the Canadian Parks Council, "Principles and Guidelines for Ecological Restoration in Canada's Protected Natural Areas," Gatineau Quebec, 2008, modified on February 3, 2011, www.pc.gc.ca/docs/pc/guide/resteco/index_e.asp (accessed February 24, 2011).

35. Thom Alcoze, "First People in the Pines: Healing the Homeland," in *Ecological Restoration of the Southwestern Ponderosa Pine Forests*, ed. Peter Friederici (Washington, DC: Island Press, 2003), 55–56. For an overview of thinking about the role of indigenous peoples in restoration, see the special issue of *Ecological Restoration* 21, no. 4 for December 2003, "Native American Land Management Practices in National Parks."

36. W. Wallace Covington, "The Evolutionary and Historical Context," in *Ecological Restoration of Southwestern Ponderosa Pine Forests*, 32–38; see also W. Wallace Covington and Margaret M. Moore, "Southwestern Ponderosa Forest Structure," *Journal of Forestry* 92 (January 1994): 39–47, for an account based on a Beaver Creek watershed south of Flagstaff.

37. Covington and Moore, "Southwestern Ponderosa Forest Structure," 45. A brief account of early forest restoration in northern Arizona is in Ecological Restoration Institute, Northern Arizona University, www.eri.nau/en/about-the-eri (accessed February 24, 2011); see also Covington, "Evolutionary and Historical Context," 46.

38. Evan Hjerpe, Jesse Abrams, and Dennis R. Becker, "Socioeconomic Barriers to the Role of Biomass Utilization in Southwestern Ponderosa Pine Restoration,"

Ecological Restoration 27 (June 2009): 169–70, 171–75; Peter Friederici, "The Flagstaff Model," in *Ecological Restoration of Southwestern Ponderosa Pine Forests*, 25.

Chapter 8: Realization III: Finding a Voice

1. Steve Packard interview with William R. Jordan III, June 5, 2009.

2. John Cairns Jr., K. L. Dickson, and E. E. Herricks, *Recovery and Restoration of Damaged Ecosystems* (Charlottesville: University of Virginia Press, 1977).

3. Peter S. White and Susan P. Bratton, "After Preservation: Philosophical and Practical Problems of Change," *Biological Conservation* 18 (1980): 241–55.

4. The proceedings of the University of Georgia symposium were published as John Cairns Jr. et al., eds., *The Recovery Process in Damaged Ecosystems* (Ann Arbor, MI: Ann Arbor Science Publishers, 1980). However, this was preceded by five years by another symposium organized by Cairns, the proceedings of which were published as John Cairns Jr., ed., *The Recovery and Restoration of Damaged Ecosystems* (Charlottesville: University Press of Virginia, 1975). John Cairns interview with William R. Jordan III, October 17, 2003.

5. "Making a User-Friendly National Park for Costa Rica: A Visit with Daniel Janzen," *Restoration & Management Notes* 5, no. 2 (Winter 1987): 72–75. Interestingly, this is the same word Aldo Leopold used in both of the written versions of his address at the UW–Madison Arboretum in 1934.

6. M. L. Heinselman, "Vegetation Management in Wilderness Areas and Primitive Parks," *Journal of Forestry* 65, no. 6 (June 1965): 440–45. Thanks to David Egan for drawing our attention to the history of use of the word *restoration* in this journal.

7. See Chris Maser's *The Redesigned Forest* (San Pedro, CA: R&E Miles, 1988) and *Ecological Diversity: The Vital and Forgotten Dimension* (Boca Raton, FL: Lewis, 1999).

8. For an account of this early defense of restoration against an old guard, see articles by Marylee Guinon in *S.E.R. News* 2, no. 2 (Summer 1989); 3, no. 1 (March 1990); and *Ecesis*, its renamed successor, 3, no. 3 (Summer 1993). Ms. Guinon, of Sycamore Associates, Berkeley, California, provided a copy of her letter to Mr. Walt, dated September 9, 1989.

9. Thanks to Emily Jones of Land and Water for providing this information.

10. Curt Meine, "Leopold's Fine Line," in *Correction Lines: Essays on Land, Leopold and Conservation* (Washington, DC: Island Press, 2004), 89–116.

11. Peter S. White, "How Do We Ensure Our Natural Area Parks Function to Preserve Species and Natural Systems?," *Natural Areas Journal* 1, no. 2 (1981): 9–10.

12. There is a pattern here. Journalists and others who observed the emerging practice of restoration from the outside have played an important role in its discovery and realization. Leslie Sauer, for example, says that the group at Andropogon noted that some of the first attention paid to this form of land management appeared in popular magazines, and a search for articles about restoration in a number of major newspapers supports her impression.

13. John J. Berger, ed., *Environmental Restoration: Science and Strategies for Restoring the Earth* (Washington, DC: Island Press, 1990). The National Research Council study was published as *Restoration of Aquatic Ecosystems: Science, Technology and Public Policy* (Washington, DC: National Academies Press, 1992).

14. E. O. Wilson, *The Diversity of Life* (Cambridge, MA: Belknap, 1992), 340.

15. Michael Soulé, quoted in "Ecologists Seek Respect, Financial Support," *Chronicle of Higher Education* 36, no. 28 (March 28, 1990): 7.

16. Bruce Babbitt, *Cities in the Wilderness: A New Vision of Land Use in America* (Washington, DC: Island Press, 2005), 53.

17. David Brower, *Work in Progress* (Layton, UT: Peregrine Smith Books, 1991), 284.

Chapter 9: Realization IV: Getting Real

1. John J. Berger, *Restoring the Earth: How Americans Are Working to Renew Our Damaged Environment* (New York: Anchor, 1985), 55–65. See also William R. Jordan III, "Hint of Green: Making Marshes along the Atlantic Coast," *Restoration and Management Notes* 1, no. 4 (1983): 4–10.

2. Berger, *Restoring the Earth*, 60.

3. Ibid., 118–22; Harold L. Nelson, "Prairie Restoration in the Chicago Area," *Restoration & Management Notes* 5, no. 2 (1987): 60–64. Cited hereafter as *R&MN*.

4. Bob Jenkins interview with William R. Jordan III, August 15, 2006.

5. Bob Jenkins interview with William R. Jordan III, August 4, 2010.

6. David J. Parsons, "Objects or Ecosystems: Giant Sequoia Management in National Parks," Symposium on Great Sequoias: Their Place in the Ecosystem and Society, June 1992, Visalia, California, www.nps.gov/seki/naturescience/fix-obj.eco.htm; National Park Service, "National Parks for the 21st Century: The Vail Agenda," Report and recommendations to the director of the National Park Service (Post Mills, VT: Chelsea Green, 1993), 105.

7. "Significant Ecological Restoration Projects Completed and Ongoing in Yosemite Valley," October 12, 2006; www.nps.gov/yose/parknews/restoration1012.htm; Yosemite National Park, "Happy Isles Fen Ecological Restoration," www.nps.gov/yos /parkmgmnt/upload/fen%5B1%50.pdf; docstoc.com.docs/6817427/Yosemite-National -Park-Volunteer-Opportunities-Projects-Starting.   www.volunteer.gov/gov/results.cfm ?ID=9496.

8. Rick Potts interview with William R. Jordan III, July 2, 2009.

9. Glenn Plumb interview with William R. Jordan III, July 8, 2009. Not surprisingly, the prospect of restoring bison to the Great Plains has inspired a good deal of serious attention to the possibility that restoration of the ecology of the region might be the basis for restoration of the economy in an area that has been beset by declining population and economics in recent decades. In *Where the Buffalo Roam: The Storm over the Revolutionary Plan to Restore America's Great Plains* (New York: Grove Weidenfeld, 1992), journalist Anne Matthews explores the response to Frank and Deborah Popper's proposal for the creation of a "Buffalo Commons" on the Great Plains. Ernest Callenbach explores the idea further in *Bring Back the Buffalo! A Sustainable Future for America's Great Plains* (Washington, DC: Island Press, 1996).

10. See Mary Doyle and Cynthia A. Drew, eds., *Large-Scale Ecosystem Restoration* (Washington, DC: Island Press, 2008), especially "The Watershed-Wide Science-Based Approach to Ecosystem Restoration" by Mary Doyle, pp. ix–xiv.

11. Other important examples of large-scale projects include the efforts of the Army Corps of Engineers to restore the greater Everglades system in southern Florida; see Michael Grunwald, *The Swamp: The Everglades, Florida and the Politics of Paradise* (New York: Simon & Schuster, 2006); and work being carried out by CALFED, an

interagency program, on behalf of restoration in San Francisco Bay and its delta; see www.dfg.ca.gov/erp.

12. Karen Rodriguez interview by William R. Jordan III, June 24, 2009.

13. See A. B. Swengel, "A Literature Review on Insect Responses to Fire, Compared to Other Conservation Management of Open Habitat," *Biodiversity and Conservation* 10, no. 7 (2001): 1141–69.

14. See articles by Owen J. Sexton et al. and William T. Leja in *Status and Conservation of Midwestern Amphibians*, ed. Michael J. Lanno (Iowa City: University of Iowa Press, 1998).

15. L. E. Frelich et al., "Earthworm Invasion into Previously Earthworm-Free Temperate and Boreal Forests," *Biological Invasions* 8 (2006): 1235–45; L. Heneghan, J. Steffen, and K. Fagen, "Interactions of an Introduced Shrub and Introduced Earthworms in an Illinois Urban Woodland: Impact on Leaf Litter Decomposition," *Pedobiologia* 50 (2006): 543–51.

16. Personal communication by Mike Redmer, U.S. Fish & Wildlife Service, Barrington, Illinois.

17. L. Heneghan et al., "Integrating a Soil Ecological Perspective into Restoration Management," *Restoration Ecology* 16, no. 4 (2008): 608–17.

18. These comments are based on observations by David Amme in an interview with W.R.J., May 29, 2009. For an overview of recent thinking about genetic considerations in restoration, see Donald J. Falk, Eric Knapp, and Edgar O. Guerrant, "An Introduction to Restoration Genetics," www.nps.gov/plants/restore/pubs/restgene/toc.htm.

19. See Joel P. Olfelt, David P. Olfelt, and Jennifer L. Ison, "Revegetation of a Trampled Cliff-Edge Using Three-Toothed Cinquefoil and Poverty Grass: A Case Study at Tettegouche State Park, Minnesota," *R&MN* 27, no. 2 (2009): 200–9; David A. Pritchett, "Vernal Pool Restoration Methods Evaluated (California)," *R&MN* 8, no. 1 (1990): n25; Terry Darby, "Cave Restoration Underway at Oregon Caves National Monument," *R&MN* 6, no. 2 (1988): n123.

20. For a survey of legislation and regulations that are broadly consistent with ecological restoration, see John J. Berger, "The Federal Mandate to Restore: Laws and Policies on Environmental Restoration," *The Environmental Professional* 13 (1991): 195–206; also Harmon Kallman et al., eds., *Restoring America's Wildlife, 1937–1987: The First 50 Years of the Federal Aid in Wildlife Restoration* (Washington: USGPO, 1987); and Michael J. Bean and Melanie J. Rowland, *The Evolution of National Wildlife Law*, 3rd ed. (Westport, CT: Praeger, 1997).

21. 16 U.S.C. § 1131–1136, 78 Stat. 890 (1964). Pub. L. No. 88-577. The Wilderness Act is presented in full in *The Great New Wilderness Debate*, ed. J. Baird Callicott and Michael P. Nelson (Athens: University of Georgia Press, 1998), 120–30.

22. J. Baird Callicott, "Explicit and Implicit Values," in *The Endangered Species Act at Thirty*, ed. J. Michael Scott, Dale D. Goble, and Frank W. Davis (Washington, DC: Island Press, 2006), Vol. 2, 39.

23. Linda Burlington interview with William R. Jordan III, September 14, 2009.

24. Linda Burlington interview; "Oil Pollution Act Overview," www.epa.gov/OEM/content/lawsreg/opaover.htm.

25. "Natural Resource Damage Assessment," www.epa.gov/superfund/programs/nrd/nrda2.htm; also restoration.doi.gov/index.html.

26. Mike Hooper interview with William R. Jordan III, July 7, 2009.

27. Matthias Gross, *Ignorance and Surprise: Science, Society, and Ecological Design* (Cambridge, MA: MIT Press, 2010), 21ff.

28. Adler came from a nursery background and also had earned a M.S. in botany from the University of Utah; see John E. Ross, "NPI," *R&MN* 3 (Summer 1985): 6–11.

29. Storm Cunningham, *The Restoration Economy: The Greatest New Growth Economy* (San Francisco: Berrett-Koehler Publishers, 2002), 1; see Barbara L. Bedford's critique in *Ecological Restoration* 21, no. 2 (June 2003): 154–55.

30. Quoted in Brian Lavendel, "The Business of Ecological Restoration," *Ecological Restoration* 20, no. 3 (2002): 174.

31. Ibid., 176.

32. André F. Clewell and James Aronson, *Ecological Restoration: Principles, Values, and Structure of an Emerging Profession* (Washington, DC: Island Press, 2007), 164.

Chapter 10: Realization V: The Relationship

1. John Cairns Jr., "Ecosocietal Restoration: Reestablishing Humanity's Relationship with Natural Systems," *Environment* 37, no. 4 (1995): 4–33; Eric S. Higgs, "What Is Good Ecological Restoration?," *Conservation Biology* 11, no. 2 (April 1997): 341, 345–47.

2. Andrew Light, "Restoration, the Value of Participation, and the Risks of Professionalization," in *Restoring Nature: Perspectives from the Social Sciences and Humanities*, ed. Paul H. Gobster and R. Bruce Hull (Washington, DC: Island Press, 2000), 163–65. Cited hereafter as *Restoring Nature*.

3. Notable examples are Stephanie Mills, *In Service of the Wild: Restoring and Reinhabiting Damaged Land* (Boston: Beacon Press, 1995), and Freeman House, *Totem Salmon: Life Lessons from Another Species* (Boston: Beacon Press, 1999). A number of writers have contributed excellent third-person accounts of the experiences of restorationists; notable examples are Bill Stevens, *Miracle under the Oaks: The Revival of Nature in America* (New York: Pocket Books, 1995), and Peter Friederici, *Nature's Restoration: People and Places on the Front Lines of Conservation* (Washington, DC: Island Press, 2006).

4. Robert E. Grese et al., "Psychological Benefits of Volunteering in Stewardship Programs," in *Restoring Nature*, 265; Herbert W. Schroeder, "The Restoration Experience: Volunteers' Motives, Values, and Concepts of Nature," in *Restoring Nature*, 247–48.

5. The essay, Packard recalls, was "What Is a Prairie?," the text accompanying the photographs in Torkel Korling, *The Prairie: Swell and Swale, from Nature* (Minneapolis: University of Minnesota Press, 1972).

6. Stevens, *Miracle under the Oaks*, 48ff.

7. Steve Packard interview with William R. Jordan III, June 5, 2009.

8. Michael Reuter interview with William R. Jordan III, June 22, 2009.

9. "Wilderness," in *A Sand County Almanac with Essays on Conservation from Round River* (New York: Ballantine, 1966), 278.

10. Pete Holloran, "The Greening of the Golden Gate," *R&MN* 14, no. 2 (Winter 1996): 112–23.

11. George E. Self, "The Human Dimension of Ecological Restoration: Attitudes of USDA Forest Service Personnel about the Role of Volunteers" (M.A. thesis, Prescott College, 2002).

12. Grese et al., "Psychological Benefits," 265; Schroeder, "The Restoration Experience," 247–48.

13. Grese et al., "Psychological Benefits," 265–66, 279; Self, "The Human Dimension of Ecological Restoration," 9–10.

14. Peter Leigh, "The Ecological Crisis, the Human Condition, and Community-Based Restoration as an Instrument for Its Cure," *Ethics in Science and Environmental Politics* (2004): 8, 11, www.globalrestorationnetwork.org/uploads/files/LiteratureAttachments/316_the-ecological-crisis-human-condition-and-community-based-restoration-as-an-instrument-for-its-cure.pdf (accessed March 3, 2011); John Thelen Steere, "Restoring Our Bonds with Place: Cultivating Commons and Community from an Eco-Psychological Perspective," paper presented at the International Conference of the Society for Ecological Restoration, Rutgers, NJ, June 20, 1996. Bill Jordan explores a number of ideas about the value of restoration for the participants in *The Sunflower Forest* (Berkeley: University of California Press, 2003).

15. Schroeder, "The Restoration Experience," 259–60. Also see Paul H. Gobster and Susan C. Barro, "Negotiating Nature: Making Restoration Happen in an Urban Park Context," pp. 185–207 in Gobster and Hull, *Restoring Nature*.

16. Robert L. Ryan, Rachel Kaplan, and Robert E. Grese, "Predicting Volunteer Commitment in Environmental Stewardship Programmes," *Journal of Environmental Planning and Management* 44, no. 5 (2001): 629–48; Irene Miles, William C. Sullivan, and Frances Kuo, "Ecological Restoration Volunteers: The Benefits of Restoration," *Urban Ecosystems* 2 (1998): 27–41; Grese et al., "Psychological Benefits," 265–80.

17. William R. Jordan III, "Environmental Junkpicking and the American Garden," *R&MN* 6, no. 2 (Winter 1988): 62; Andrew Light, "Restoration, the Value of Participation, and the Risks of Professionalization," pp. 163–81 in Gobster and Hull, *Restoring Nature*.

18. David M. Ostergren, Jesse B. Abrams, and Kimberly A. Lowe, "Fire in the Forest: Public Perceptions of Ecological Restoration in North-Central Arizona," *Ecological Restoration* 26 (March 2008): 51; Peter Friederici, "The 'Flagstaff' Model," in *Ecological Restoration of Southwestern Ponderosa Pine Forests*, ed. Peter Friederici (Washington, DC: Island Press, 2003), 25.

19. Susan C. Barro and Alan D. Bright, "Public Views on Ecological Restoration," *R&MN* 16, no. 1 (Summer 1998): 59–65.

20. Debra Shore, "Controversy Erupts over Restoration in Chicago Area," *R&MN* 15, no. 1 (Summer 1997): 25–31.

21. For reflective overviews see Debra Shore, "Controversy Erupts over Restoration in Chicago Area," *R&MN* 15, no. 1 (Summer 1997): 25–31; Alf Siewers, "Making the Quantum-Culture Leap," *R&MN* 16, no. 1 (1998): 9–15; and Reid Helford, "Constructing Nature as Constructing Science: Expertise, Activist Science and Public Conflict in the Chicago Wilderness," in *Restoring Nature*, 119–42.

22. For discussions of the "Chicago controversy," see articles by Laurel Ross, Debra Shore, and Paul H. Gobster in *R&MN* 15, no. 1 (1997); by Alf Siewers in *R&MN* 15, no. 1 (1998); by Paul H. Gobster (Introduction), R. Bruce Hull et al. (chapter 5), Reid M. Helford (chapter 6), Joanne Vining et al. (chapter 7), and Andrew Light (chapter 8) in *Restoring Nature*; and reviews of this book by William Throop, Harry M. Webne-Behrman, Karen M. Rodriguez and Kent Fuller, and Paul H. Gobster and R. Bruce Hull in *R&MN* 19, no. 4 (2001). For debates over restoration within the

Chicago conservation community, see Jon Mendelson et al., "Carving Up the Woods," *R&MN* 10, no. 2 (Winter 1992): 127–31; and Steve Packard, "Restoring Oak Ecosystems," *R&MN* 11, no. 1 (Summer 1993): 5–16.

23. Helford, "Constructing Nature," 138.

24. Evan Hjerpe, Jesse Abrams, and Dennis R. Becker, "Socioeconomic Barriers and the Role of Biomass Utilization in Southwestern Ponderosa Pine Restoration," *Ecological Restoration* 27 (June 2009): 170–72.

25. Chris Woodall, Allison Handler, and Len Broberg, "Social Dilemmas in Grassland Ecosystem Restoration: Integrating Ecology and Community on a Montana Mountainside," *Ecological Restoration* 18 (Spring 2000): 39.

26. Ibid., 40–44.

27. Marilyn J. Marler et al., "Changing Attitudes about Grassland Conservation in Missoula, Montana: 'Weed Capital of the West,'" *Ecological Restoration* 23 (March 2005): 29–34.

28. Ibid., 30–34.

29. Kimberly Bosworth Phalen, "An Invitation for Public Participation in Ecological Restoration: The Reasonable Person Model," *Ecological Restoration* 27 (June 2009): 179–80.

30. Paul Gobster, "Visions of Nature: Conflict and Compatibility in Urban Park Restoration," *Landscape and Urban Planning* 56 (2001): 35–51.

31. Helford, "Constructing Nature," 122.

32. A. Brownlow, "Inherited Fragmentations and Narratives of Environmental Control in Entrepreneurial Philadelphia," in *In the Nature of Cities*, ed. N. Heynen, M. Kaika, and E. Swyngedouw (New York: Routledge, 2006), 208–25.

33. J. T. Curtis and M. L. Partch, "Effect of Fire on the Competition between Bluegrass and Certain Prairie Plants," *The American Midland Naturalist* 39, no. 2 (1948): 237–43.

34. Ibid., 6. The question of the *locus classicus* of the term *restoration ecology* is complicated by the fact that Jordan and Aber first used it to refer to the practice of ecological restoration generally, in J. D. Aber and W. R. Jordan III, "Restoration Ecology: An Environmental Middle Ground," *BioScience* 35 (1985): 399. The next year they defined it in a quite different way, as "restoration (and ecological management generally) a technique for raising basic questions and testing fundamental ideas, leading in turn to improved restoration and management techniques."

35. William R. Jordan III, Michael P. Gilpin, and John D. Aber, "Restoration Ecology: Ecological Restoration as a Technique for Basic Research," in *Restoration Ecology: A Synthetic Approach to Ecological Research*, ed. W. R. Jordan et al. (Cambridge: Cambridge University Press, 1987), 3–21.

36. Charles Darwin, *The Origin of Species* (New York: Modern Library, n.d.), 55. Also relevant is Thoreau's discussion of experiments on the use of nurse plants in establishing stands of oaks. Note especially his comment that "when we experiment in planting forests, we find ourselves at last doing as Nature does. Would it not be well to consult with Nature in the outset? For she is the most extensive and experienced planter of us all." Henry David Thoreau, "The Succession of Forest Trees," in *Thoreau's Complete Works* (Cambridge, MA: Riverside Press, 1906), 184–204.

37. A. D. Bradshaw, "The Reconstruction of Ecosystems," *Journal of Applied Ecology* 20 (1983): 1–17.

38. Joy Zedler interview with William R. Jordan III, July 16, 2009.

39. "Restoration Ecology," uwarboretum.org/research/study/; see also Joy B. Zedler, "Where Restoration Emerged: A Gem of a Reserve Celebrates Its Diamond Anniversary," *Ecological Restoration* 27 (June 2009): 109–10.

40. "River Restoration at Berkeley," University of California, www.lib.berkeley .edu/WRCA/restoration/index.html; also   www.lib.berkeley.edu/WRCA/restoration /rcourses.html (accessed May 15, 2009 and May 26, 2009).

41. Not all institutions advertise their ecological restoration programs in online catalogs, nor do academic emphases on the discipline consistently show up in undergraduate or graduate catalogs. Consequently, a comprehensive survey of institutions offering training in ecological restoration is almost impossible.

42. www.bae.ncsu.edu/programs/extension/wqg/srp/; www.clemson.edu/restoration /focus_areas/restoration_ecology/projects/sand_river.

43. Peter Friederici, ed., *Ecological Restoration of Southwestern Ponderosa Pine Forests* (Washington, DC: Island Press, 2003), xix–xx; Truman Young, personal communication, September 22, 2009; Peter Fulé interview with G.M.L., August 24, 2009; and Northern Arizona University Undergraduate and Graduate Catalogs, 2010–2011, www4.nau.edu/academiccatalog2010/academiccatalog.htm (accessed March 3, 2011).

44. Dan Binkley, personal correspondence with William R. Jordan III, September 21, 2008.

45. William Jordan III explores the role of ritual in the creation of value and its relevance for the development of restoration at some length in *The Sunflower Forest*. See especially pp. 46–53 and chapters 6 and 7.

46. Catherine Pickstock, *After Writing: The Liturgical Consummation of Philosophy* (Oxford: Blackwell, 1998), cited in Jordan, *The Sunflower Forest*, 142, 186.

47. Pete Holloran interview with William R. Jordan III, June 3, 2009.

48. It is interesting to note that although the Leopold Report on management of the national parks cautioned against showcasing of management operations in order to preserve an illusion of naturalness (Jordan, *Sunflower Forest*, 161–62), James Pritchard notes that in a talk he gave at about the time the report came out, Starker Leopold himself urged that because the public viewed the parks mainly from roads, "roadside wildlife management may be the most important facet of all in creating this vision I am talking about." James A. Pritchard, *Preserving Yellowstone's Natural Conditions* (Lincoln: University of Nebraska Press, 1999), 213.

49. Eric Higgs, *Nature by Design: People, Natural Process and Ecological Restoration* (Cambridge, MA: MIT Press, 2003), 185ff.

50. Ibid.

51. Ibid., 225–59. Also see Lisa Meekison and Eric Higgs, "The Rites of Spring (and Other Seasons): The Ritualization of Restoration," *R&MN* 16, no. 1 (1998): 73–81.

52. C. Christopher Norden, "Beyond Apocalypse: Some Reflections on Native Writers, Modern Literature—And Restoration and Ritual," *R&MN* 11, no. 1: 45–51.

53. Michael L. Rosenzweig, *Win–Win Ecology: How the Earth's Species Can Survive in the Midst of Human Enterprise* (Oxford: Oxford University Press, 2003), 70ff.

54. Karen M. Holland, "Restoration Rituals: Transforming Workday Rituals into Inspirational Moments," *R&MN* 12, no. 2, 121–25.

55. A short video of the April 2009 burn, narrated by Woodson, is available online at http://ssm.nwherald.com/video/926/prairie-burn/.

56. Dave Simpson, "Picks, Shovels and Musical Comedy," *R&MN* 15, no. 2 (1997): 180.

57. Freeman House, *Totem Salmon: Life Lessons from Another Species* (Boston: Beacon, 1999), 101–4.

Chapter 11: Current Thinking

1. Pauline Drobney interview with William R. Jordan III, June 15, 2009.

2. See Keith Kloor, "Returning America's Forests to their 'Natural' Roots," *Science* 287 (January 28, 2005): 573–75.

3. Mark A. David and Lawrence B. Slobodkin, "Restoration Ecology: The Challenge of Social Values and Expectations" (Forum), *Frontiers in Ecology and the Environment* 2, no. 1 (February 2004): 44–45.

4. Mark A. Davis and Lawrence B. Slobodkin, "The Science and Values of Restoration Ecology," *Restoration Ecology* 121 (2004): 1–3.

5. Thomas B. Simpson, "The Dechannelizing of Nippersink Creek: Learning about Native Illinois Streams through Restoration," *Ecological Restoration* 26, no. 4 (December 2006): 350–56.

6. Ed Collins, personal communication to William R. Jordan III, October 23, 2007.

7. Anthony D. Bradshaw, "Underlying Principles of Restoration," *Canadian Journal of Fisheries and Aquatic Science* 53, suppl. 1 (1996): 3–9.

8. Ibid.; "What Do We Mean by Restoration?," in *Restoration Ecology and Sustainable Development*, ed. Krystyna M. Rubanska, Nigel R. Webb, and Peter J. Edwards (Cambridge: Cambridge University Press, 1997), 8–14. The material quoted is on p. 10. The study, chaired by John Cairns Jr., was published as National Research Council, *Restoration of Aquatic Ecosystems: Science, Technology and Public Policy* (Washington, DC: National Academies Press, 1992).

9. J. Aronson, S. Dhillion, and E. Le Floc'h, "On the Need to Select an Ecosystem of Reference, However Imperfect: A Reply to Pickett and Parker," *Restoration Ecology* 3 (1995): 1–3. The authors are responding to S. T. A. Pickett and V. T. Parker, "Avoiding the Old Pitfalls: Opportunities in a New Discipline," *Restoration Ecology* 2 (1994): 70–75.

10. Eric Higgs, *Nature by Design: People, Natural Process and Ecological Restoration* (Cambridge, MA: MIT Press, 2003), 131ff.

11. Dave Egan and Evelyn A. Howell, eds., *The Historical Ecology Handbook: A Restorationist's Guide to Reference Ecosystems* (Washington, DC: Island Press, 2001).

12. Marcus Hall, for example, writes in *Earth Repair* that "the restorationist puts human interests first by the very attempt to exclude those interests" (196).

13. Reed F. Noss and Michael Scott, "Ecosystem Protection and Restoration: The Core of Ecosystem Management," in *Ecosystem Management: Applications for Sustainable Forest and Wildlife Resources*, ed. Mark S. Boyce and Alan Haney (New Haven, CT: Yale University Press, 1997), 239–64.

14. Michael J. Stevenson, "Problems with Natural Capital: A Response to Clewell," *Restoration Ecology* 8, no. 3 (2000): 211–13.

15. David Lowenthal, "Reflections on Humpty-Dumpty Ecology," in *Restoration*

and History: The Search for a Useable Environmental Past, ed. Marcus Hall (New York: Routledge, 2010), 13–34. The study Lowenthal cites is David A. Sear and Jim A. Milne, "Does Scientific Conjecture Accurately Describe Restoration Practice? Insight from an International River Restoration Survey," *Area* 38 (2006): 128–42.

16. James Aronson, Suzanne J. Milton, and James N. Blignaut, "Restoring Natural Capital: Definitions and Rationale," in *Restoring Natural Capital: Science, Business and Practice*, ed. James Aronson, Suzanne J. Milton, and James N. Blignaut (Washington, DC: Island Press, 2007), 3–8. The item quoted is on p. 5; Andre F. Clewell and James Aronson, *Ecological Restoration: Principles, Values, and Structure of an Emerging Profession* (Washington, DC: Island Press, 2007), especially chapters 7 and 8. For an overview of ideas about the economic value of ecosystems, see Gretchen Daily, *Nature's Services: Societal Dependence on Natural Ecosystems* (Washington, DC: Island Press, 1997).

17. Andre F. Clewell, "Restoration of Natural Capital," *Restoration Ecology* 8, no. 1 (2000): 1–2.

18. Ibid., 56.

19. K. G. Lyons et al., "Rare Species and Ecosystem Functioning," *Conservation Biology* 19, no. 4 (2005), 1–6; M. W. Schwartz et al., "Linking Biodiversity to Ecosystem Function: Implications for Conservation Biology," *Oecologia* 122: (2000): 297–305; Mark Schwartz, interview by William R. Jordan III, January 29, 2010.

20. David Nirenberg, "Love and Capitalism" (review of an encyclical by Pope Benedict XVI), *The New Republic* 240, no. 4868 (September 23, 2009).

21. Nathaniel E. Seavy et al., "Why Climate Change Makes Riparian Restoration More Important Than Ever," *Ecological Restoration* 27, no. 3 (Spring 2009): 330; James A. Harris et al., "Ecological Restoration and Global Climate Change," *Ecological Restoration* 14, no. 2 (June 2006): 171. Also see Douglas Fox, "Back to the No-Analog Future?," *Science* 316 (May 11, 2007): 823–25.

22. Harris et al., "Ecological Restoration and Global Climate Change," 173; Constance I. Millar, Nathan L. Stephenson, and Scott L. Stephens, "Climate Change and Forests of the Future: Managing in the Face of Uncertainty," *Ecological Applications* 17, no. 8 (2007), 2150. For an early expression of this interest see Robert L. Peters II, "Global Climate Change: A Challenge for Restoration Ecology," *Restoration and Management Notes* 3, no. 2 (1985): 62–67.

23. Stephen B. Glass, Bradley M. Herrick, and Christopher J. Kucharik, "Climate Change and Ecological Restoration at the University of Wisconsin–Madison Arboretum," *Ecological Restoration* 27, no. 3 (September 2009): 348. This is one of five articles on restoration and climate change in this issue of *Ecological Restoration*.

24. Seavy et al., "Why Climate Change Makes Riparian Restoration More Important Than Ever," 333–36.

25. Constance I. Millar, Nathan L. Stephenson, and Scott L. Stephens, "Climate Change and Forests of the Future: Managing in the Face of Uncertainty," *Ecological Applications* 17, no. 8 (2007): 2145–46.

26. Ibid., 2150.

27. Peter W. Dunwiddie et al., "Rethinking Conservation Practice in Light of Climate Change," *Ecological Restoration* 27, no. 3 (September 2009): 321.

28. Liam Heneghan, personal communication to William R. Jordan III, November 17, 2010.

29. R. J. Hobbs and J. A. Harris, "Restoration Ecology: Repairing the Earth's Ecosystems in the New Millennium," *Restoration Ecology* 9 (2001): 239–46; Richard J.

Hobbs et al., "Novel Ecosystems: Theoretical and Management Aspects of the New Ecological Order," *Global Ecology and Biogeography* 15: (2006): 1–7; Timothy R. Seastedt, Richard J. Hobbs, and Katharine N. Suding, "Management of Novel Ecosystems: Are Novel Approaches Required?," *Frontiers in Ecology and the Environment* 6, no. 10 (2008): 547–53.

30. See Dario Papale and Riccardo Valentini, "The Global Carbon Cycle: Current Research and Uncertainties in the Sources and Sinks of Carbon," in *Ecological Restoration: A Global Challenge*, ed. Francisco A. Comin (Cambridge: Cambridge University Press, 2010), 21–44.

31. For an example of the misgivings aroused by the suggestion that disciplines outside the natural sciences and areas of experience other than the technical may have important roles to play in restoration, see Eric Higgs, "Expanding the Scope of Ecological Restoration," *Ecological Restoration* 2, no. 3 (1994): 137–46, and Anthony Bradshaw's response, "The Need for Good Science: Beware of Straw Men: Some Answer to Comments by Eric Higgs," 147–48.

32. Jan E. Dizard, "Going Native: Second Thoughts about Restoration," in *Reconstructing Conservation: Finding Common Ground*, ed. Ben A. Minter and Robert E. Manning (Washington, DC: Island Press, 2003), 43–56.

33. Thomas B. Simpson, "Old Nature, New Nature: Global Warming and Restoration," *Ecological Restoration* 27, no. 3 (September 2009): 343–44; Fox, "Back to the No-Analog Future?," 823–25.

34. See the exchange between J. Aronson and his colleagues and Bill Jordan, "Is *Ecological Restoration* a Journal for North American Readers Only?," *Ecological Restoration* 18, no. 3 (2000): 146–49.

35. Society for Ecological Restoration International, Science and Policy Working Group, *The Society for Ecological Restoration International Primer on Ecological Restoration*, 2004. www.ser.org/content/ecological_restoration_primer.asp.

36. Ralph Waldo Emerson, "Nature," in *Essays: Second Series*, vol. III of *Emerson's Complete Works* (Boston: Houghton Mifflin, 1983), 161–88. The item quoted is on p. 182.

THE SCIENCE AND PRACTICE
OF ECOLOGICAL RESTORATION

Wildlife Restoration: Techniques for Habitat Analysis and Animal Monitoring, by Michael L. Morrison

Ecological Restoration of Southwestern Ponderosa Pine Forests, edited by Peter Friederici, Ecological Restoration Institute at Northern Arizona University

Ex Situ Plant Conservation: Supporting Species Survival in the Wild, edited by Edward O. Guerrant Jr., Kayri Havens, and Mike Maunder

Great Basin Riparian Ecosystems: Ecology, Management, and Restoration, edited by Jeanne C. Chambers and Jerry R. Miller

Assembly Rules and Restoration Ecology: Bridging the Gap between Theory and Practice, edited by Vicky M. Temperton, Richard J. Hobbs, Tim Nuttle, and Stefan Halle

The Tallgrass Restoration Handbook: For Prairies, Savannas, and Woodlands, edited by Stephen Packard and Cornelia F. Mutel

The Historical Ecology Handbook: A Restorationist's Guide to Reference Ecosystems, edited by Dave Egan and Evelyn A. Howell

Foundations of Restoration Ecology: The Science and Practice of Ecological Restoration, edited by Donald A. Falk, Margaret A. Palmer, and Joy B. Zedler

Restoring the Pacific Northwest: The Art and Science of Ecological Restoration in Cascadia, edited by Dean Apostol and Marcia Sinclair

A Guide for Desert and Dryland Restoration: New Hope for Arid Lands, by David A. Bainbridge

Restoring Natural Capital: Science, Business, and Practice, edited by James Aronson, Suzanne J. Milton, and James N. Blignaut

Old Fields: Dynamics and Restoration of Abandoned Farmland, edited by Viki A. Cramer and Richard J. Hobbs

Ecological Restoration: Principles, Values, and Structure of an Emerging Profession, by Andre F. Clewell and James Aronson

River Futures: An Integrative Scientific Approach to River Repair, edited by Gary J. Brierley and Kirstie A. Fryirs

Large-Scale Ecosystem Restoration: Five Case Studies from the United States, edited by Mary Doyle and Cynthia A. Drew

New Models for Ecosystem Dynamics and Restoration, edited by Richard J. Hobbs and Katharine N. Suding

Cork Oak Woodlands on the Edge: Ecology, Adaptive Management, and Restoration, edited by James Aronson, João S. Pereira, and Juli G. Pausas

Restoring Wildlife: Ecological Concepts and Practical Applications, by Michael L. Morrison

Restoring Ecological Health to Your Land, by Steven I. Apfelbaum and Alan W. Haney

Restoring Disturbed Landscapes: Putting Principles into Practice, by David J. Tongway and John A. Ludwig

Intelligent Tinkering: Bridging the Gap between Science and Practice, by Robert J. Cabin

Making Nature Whole: A History of Ecological Restoration, by William R. Jordan and George M. Lubick